Coevolution of Animals and Plants

The Dan Danciger Publication Series

Coevolution
of
Animals and Plants

Symposium V
First International Congress of
Systematic and Evolutionary
Biology
Boulder, Colorado
August 1973

Edited by

Lawrence E. Gilbert
and

Peter H. Raven

UNIVERSITY OF TEXAS PRESS, AUSTIN AND LONDON

For reasons of economy and speed this
volume has been printed from camera-
ready copy furnished by the Editors,
who assume full responsibility for its
contents.

International Standard Book Number 0-292-71031-3
Library of Congress Catalog Card Number 74-11787
Copyright © 1975 by the University of Texas Press
Printed in the United States of America

Contents

GENERAL INTRODUCTION

Lawrence E. Gilbert

Department of Zoology
The University of Texas
Austin, Texas 78712

Peter H. Raven

Missouri Botanical Garden
2315 Tower Grove Avenue
St. Louis, Missouri 73110

The papers in this collection deal in a variety of ways with the coevolution of plants and animals, and with the specification of the ecological interactions between them. Although it has long been recognized that plants and animals profoundly affect one another's characteristics during the course of evolution, the importance of coevolution as a dynamic process involving such diverse factors as chemical communication, population structure and dynamics, energetics, and the evolution, structure, and functioning of ecosystems has been recognized widely for less than a decade. Indeed, coevolution really represents a point of view about the structure of nature which has just begun to be fully and rigorously exploited, and which holds great promise for the future.

These papers were presented initially as a symposium on the coevolution of animals and plants at the First International Congress of Systematic and Evolutionary Biology, Boulder, Colorado, 10-11 August 1973. They have been revised subsequently, in some cases rather extensively. Miriam Rothschild was unfortunately unable to attend the symposium owing to last-minute complications, so that her stimulating paper is presented here for the first time. Brian Hocking, one of the participants in the symposium, died suddenly a few months later, 23rd May, 1974, but had submitted his written paper previously. We were priveleged to hear the fine contribution by this pioneer in the energetics of insect flight, and are fortunate to have the opportunity to publish it here. Dr. Hocking's incisive mind and stimulating personality will be missed by his many colleagues, as they are by us.

Insofar as we are aware, there is no other volume focusing on the dynamic aspects of animal-plant coevolution. In selecting the participants for this symposium, we attempted to cover, as broadly as possible, all of the ways in which plants interact with animals. We therefore have, for example, included papers which deal with leaf-feeding animals and their impact on plant evolution (Feeny) as well as those which concentrate on predator-prey relationships involving the seeds of angiosperms (Smith). Several papers deal with mutualistic relationships between plants and animals: Hocking deals with ant-plant relationships in a broad context; McKey deals with the mutualistic aspects of seed dispersal and points the way to interesting models that can be constructed in this area. Several authors deal with the most familiar aspect of mutualistic plant-animals interactions, namely pollination relationships. Finally, Rothschild provides a fascinating example of more indirect relationships centered around the role of carotenoids which are produced by plants but play such a fundamental role in visual systems both in plants and in animals.

In addition, within the broad area of population biology, we attempted to include the greatest possible diversity of subjects and levels of approach. Thus the papers of Feeny and Baker deal extensively with biochemistry and nutrition; Heinrich and Smith are concerned with energetics and behavior, particularly with reference to foraging strategies; and Dodson points out important relationships between the factors important in insect courtship and those involved in plant pollination. The broad patterns of community ecology, as assayed by pollination and seed-dispersal systems are treated by Frankie. Gilbert, using a carefully selected plant-insect system, has attempted to integrate examples at several levels and to provide an index to community evolution and function *via* a different route.

In general, we made an effort to encourage contributions which emphasized both plants and animals as important elements in interactive, coevolutionary systems. An emphasis was placed on studies that stressed the process rather than the products of coevolution. Several of the papers in this collection have attempted the construction of testable models and theory upon which further progress can be based, a difficult process owing to the overall complexity of the field, but a potentially rewarding one.

It should be obvious from an inspection of this volume that the papers included in it differ widely in their basic character. Several papers, for example those by McKey, Smith, and Rothschild, are reviews of areas included in the field, with new interpretations. Others deal more directly with particular experimental results, like the paper by Heinrich on the role of energetics in bumblebee-flower relationships, and that by Frankie on tropical forest ecology in Costa Rica. Some cover insects broadly, whereas others are focused on specific groups: euglossine bees, ants, *Heliconius*. Such diversity will, we hope, add interest to the overall volume.

All of the papers in this volume deal with terrestrial seed plants, and, not unnaturally, they stress the relationships of these plants with insects, the most important class of herbivores. The most important question for the future concerns the way in which relationships of the sort discussed here can be assembled into broad and therefore useful generalities. Rigorous mathematical models in the field of coevolution are inherently difficult

to construct not only because of the complexity in factoring out
trophic levels, but also because of the enormous difficulties associated with integrating ecological time and evolutionary time.
Coevolutionary relationships are by definition the product of historical change (= evolution); yet this historical change is still
proceeding. The relationships between elements in a contemporary
ecosystem are dynamic both in an evolutionary and in an ecological sense, and, depending upon the generation times involved, may
even change very rapidly during a given period of observation.
This is all the more likely in these times of extensive environmental pollution and the consequent alteration of relationships within ecosystems that must affect all of our observations.

Tightly structured organism-organism relationships hold considerable promise for understanding the area as a whole. Thus in his
treatment of ant-plant mutualism, Hocking discusses the role of
plants as food for ants and as shelter for ants, and in turn the
role of ants as protectors of plants, as pollinators, as disseminators of plant seeds, and even as food for carnivorous plants. He
stresses the role of energetics in plant-insect interactions in
general, using examples from nectar-feeding and blood-sucking groups
in arctic regions.

Heinrich also stresses the role of energy as the currency of
plant-animal coevolutionary relationships in his discussion of bumblebee-flower interactions, a theme that runs through most of the
papers in this volume. By considering the relationships of bumblebees with flowers in one ecological arena, he concludes that in
the situation he studied in Maine, synchronous blooming and colonial
occurrence were necessitated by the relatively low caloric reward
of most of the plant species.

Gilbert uses the mutualistic relationship between butterflies
of the genus *Heliconius* and plants from which they obtain their
pollen and nectar, members of the cucurbit genera *Anguria* and
Gurania, to explain the highly reticulate evolutionary system of
which they are members. The model he presents on p. 221 provides
a clear example of the complexity of such systems, but also the
promise of studies of this nature for illuminating the mode of
operation of ecological and evolutionary relationships at the ecosystem level.

Dodson considers the role of fragrance compounds in orchids,
but also in aroids and gesneriads, in attracting and maintaining
constancy by male euglossine bees. These bees, like the butterflies of the genus *Heliconius* studied by Gilbert, visit long-
blooming plants at rare intervals. Particularly valuable is Dodson's demonstration of variation in the attractiveness of certain
compounds to certain bee species, both within and between populations. This points the way to the mode of evolution of such systems.

Rothschild has considered in elegant detail the role of carotenoids in ecosystems, as visual signals in plants and as part of
the visual apparatus and also as visual signals in animals. Carotenoids mediate the flow of energy through ecosystems and play a
role in structuring this flow in much the same way that the terpenoids studied by Dodson and his associates do, but on a much more
global scale.

Herbert and Irene Baker have provided a useful early paper
in the unfolding story of the nature and quantity of constituents present in the nectar of engiosperm flowers, until recently
generally thought to be an energy-rich but rather uninteresting
solution of various sugars. The roles of such compounds as amino
acids, lipids, antioxidants, and alkaloids in nectar are only beginning to be appreciated, but it appears almost certain that the
amino acids and lipids play a nutritional role comparable with that
of the sugars. The Bakers also demonstrate differences in these
constituents that appear to be related to the normal visitors to
the flowers of the plants concerned.

The paper by Frankie is unique in this volume in emphasizing
relationships at a community level concerned with flowering phenology and broad pollinator relationships. Such widescale studies
will be increasingly necessary to provide a framework for the more
narrowly oriented and circumscribed sorts of studies represented
by most of the papers in this volume. Generalities have been possible, for example, concerning the annual cycles of occurrence of
moth-pollinated flowers and moths, and of bee-pollinated flowers
and bees, in the seasonally dry forest of Costa Rica.

The remaining three papers, dealing with herbivory (Feeny),
seed predation (Smith), and seed dispersal (McKey), may be regarded as the beginning of a process of generalization across ecosystems that will eventually make possible effective and useful model-
building in the area of coevolutionary studies. Such studies are
absolutely fundamental to an understanding of ecosystems, the relationships between plants and their herbivores structure all
higher-level interactions in ecosystems, since a far greater quantity of energy is transferred at this level than at any other.
In addition, the chemical defenses and attractants that the plants
manufacture are often utilized in other interactions involving the
herbivores that obtain these molecules from the plants; these herbivores are often incapable of manufacturing the molecules themselves.

In part, then, the relationships within ecosystems obviously
can be described most exactly in terms of an accurate specification of the pathways by which energy flows through these ecosystems,
and the dimensions of these pathways. The ecosystem as a whole
has a stability which comes about by virtue of the interacting strategies --- morphological, chemical, behavioral ones, for example ---
of its constituent organisms to alter these pathways either quali-

tatively or quantitatively. Coevolutionary studies, however, are
concerned with the dynamics of the evolutionary relationships that
have led to a given situation and to the reciprocal modifications
that have taken place in the participating organisms. Such interactions are theoretically best expressed in terms of the models of
ecology and population genetics. If this symposium is repeated
at the Second International Congress of Systematic and Evolutionary
Biology in 1979, we would expect to find much progress in the construction and testing of such models and in their intelligent application to the sorts of situations discussed in the present volume.

In conclusion, we would like to express our gratitude to the
organizers of the First International Congress of Systematic and
Evolutionary Biology for the opportunity to present this symposium
there; to the University of Texas Press for their expert attention
to this volume; to Ann Reynolds and Karla Cooper who had the difficult task of typing the total symposium in photo-ready form; to
the National Science Foundation for grants awarded individually
to each of us which have facilitated our studies in and consequently
our understanding of the field of coevolution.

Coevolution of Animals and Plants

BIOCHEMICAL COEVOLUTION BETWEEN PLANTS AND THEIR INSECT HERBIVORES

Paul Feeny

Department of Entomology and Section
of Ecology and Systematics
Cornell University
Ithaca, N. Y. 14850

INTRODUCTION

At all stages of their life histories populations of herbivorous insects and the plants upon which they feed are subject to
a host of sources of mortality, ranging from unfavorable weather
to attack by many and various pathogens, parasites and predators.
Such sources of mortality act also, of course, as selective
pressures, and every insect or plant population may be said to
coevolve to a greater or lesser extent with everything around it
that is capable of evolution. Our reasons for singling out for
discussion the coevolution between insects and plants follow from
a conspicuous non-event. The enormous destructive potential of
herbivorous insects, illustrated dramatically by occasional outbreaks of agricultural and forest pests, has failed to prevent
green plants from dominating most of the terrestrial surface of
this planet. While plants doubtless benefit to some extent from
depredations on herbivorous insects by their own natural enemies
of various kinds (Hairston, Smith and Slobodkin, 1960), there is
good reason to believe that the survival of plants is due largely to their own defensive strategies, coevolving through time in
relation to the attack strategies of herbivores and pathogens.
The reasons for suspecting that coevolution between insects
and plants has a substantial *biochemical* component stem from
three categories of observation. First, many insect species discriminate between their host plants and non-host plants in large
part by differential behavioral responses to various secondary
substances, such as alkaloids, terpenoids and mustard oil glucosides, which are of scattered distribution through the plant
kingdom (Verschaffelt, 1911; Dethier, 1947, 1970; Fraenkel, 1959;
Schoonhoven, 1972). Second, many of the secondary substances
present in plants are known to be poisonous to animals, including
insects (Stahl, 1888; Fraenkel, 1959). Third, the various food
plants of a particular insect species, genus or even family often
share similar secondary substances, even though they may differ
widely in other respects (Ehrlich and Raven, 1965).
From these basic observations there has developed over the
course of this century an hypothesis, most explicitly stated by

Ehrlich and Raven (1965) for the butterflies and their food
plants, which attributes to plant secondary compounds a key role
in determining the pattern of insect-plant coevolution. According to this hypothesis, now generally accepted, at least some of
the secondary substances found in plants were evolved or elaborated in response to attack by insects. Some of the associated
insects evolved methods of tolerating the new plant chemicals
and were thus able to remain associated with their particular
host species. As plants evolved through time, some of their
associated insects coevolved with them, often leaving relatives
on the original plant families or genera. One man's meat became
another man's poison and present-day insects are adapted to tolerate only a certain range of chemicals and therefore only a
certain range of plants. Evolving along with detoxication mechanisms, chemosensory systems responding differentially to secondary compounds enabled insects to locate the plants to which they
were chemically adapted. We are thus witnessing an evolutionary
arms race in which the plants, for survival, must deploy a fraction of their metabolic budgets on defense (physical as well as
chemical) and the insects must devote a portion of their assimilated energy and nutrients on various devices for host location
and attack.
      It has to be admitted that the experimental evidence for a
defensive function of plant secondary compounds against ecologically appropriate insects is rather weak thus far, for a number
of reasons. First, chemical defenses of plants are generally not
as conspicuous as those, for example, of many insects (Eisner,
1970). Second, plants (like some animals) may use for protection
substances of relatively subtle effect (Eisner, 1970); these may
not even be immediately toxic to an invading insect individual
but nevertheless can reduce its fitness. This serves the functions both of minimizing a population build-up by an attacker and
of selecting against those invaders which attempt to colonize the
plant. Thus a plant chemical which does not prove immediately
toxic to a bioassay insect may nevertheless represent a significant defense against insects. By contrast, since secondary chemicals serve a great variety of functions in plants (Whittaker and
Feeny, 1971), demonstration that a particular compound is toxic
to one or more insects does not necessarily indicate that this is
its ecological function. A third problem is the common difficulty of inducing a bioassay insect to feed on an abnormal diet or
host plant due to the presence of feeding inhibitors or absence
of feeding stimulants; it may thus be difficult to distinguish
between gradual toxic effects and those due to relative starvation. Removal of maxillary chemoreceptors (Waldbauer, 1962) or
rearing larvae on normal host plants which have been cultured in
solutions of the suspected toxin (Erickson and Feeny, 1974) are

possible ways of avoiding this problem. Finally, there exists
the possibility that herbivorous insects may rely in part on gut
microorganisms for detoxication of potential toxins. Should this
prove to be the case, bioassay results obtained by using diets
containing antibiotics may have to be interpreted with caution.

In spite of such problems the overall hypothesis of biochemical coevolution remains overwhelmingly convincing (see also
Brower and Brower, 1964; Fraenkel, 1969; Dethier, 1970; Whittaker
and Feeny, 1971; Levin, 1971; Rothschild, 1972a; Schoonhoven
1972; Southwood, 1972). Yet important though biochemical evolution may be as a major determinant of the evolution of phytophagous insects, it cannot explain completely the observed patterns
of insect feeding diversity. Phytophagous insects are subject to
a wide variety of selective pressures which may have little to do
directly with the chemistry of their food plants and it is the
interaction between the channelling effect of biochemical coevolution and the adaptive flexibility to respond to ecological
selective pressures which must account for the observed patterns
of insect feeding diversity. Here I would like to explore the
relative importance of these two classes of phenomena with particular reference to the Lepidoptera and their food plants and
speculate that it may be dependent to a significant extent on
plant community structure and diversity.

Herbivorous insects display a great range of feeding strategies varying from monophagous species which may attack only one
species of plant to polyphagous species which attack a great
variety of plant families in nature (Brues, 1920; Dethier, 1947).
One way of introducing a discussion about the interaction between coevolutionary determination and adaptive flexibility is
simply to ask why specialist insects tend to remain specialized.
There are two categories of answer to this question:

1. They remain specialized because even though they have
many evolutionary opportunities to extend their host plant
ranges it is not ecologically advantageous to exploit them.
2. They remain specialists because even though it might be
advantageous to extend their host ranges they rarely if ever
have the evolutionary opportunity to do so.

ECOLOGICAL ADVANTAGES OF FEEDING SPECIALIZATION

Extreme feeding specialization may allow close adaptation to
the microhabitat and phenology of a particular species of plant
though this advantage is likely to be reduced to the extent that
closely related plants within the insect's host range differ in
phenology and microhabitat. Specialization is likely, also to
permit finer adaptation to other components of the fauna associated with the host plants; thus patterns of crypsis may be more

effective and, where insects can assimilate and store toxic chemicals from their plant food, protection against predators may be enhanced (Brower, Brower and Corvino, 1967; Rothschild, 1972a and 1972b). Although a specialist feeder may be dependent on the ecological and evolutionary fate of very few plant species and may have to possess elaborate behavioral mechanisms to allow it to find these plants, it may face reduced competition from other insect species within its specialized "adaptive zone" (Dethier, 1954; Ehrlich and Raven, 1965).

Specialization may result also in appreciable reduction in metabolic costs, for example in tolerating defensive chemicals present in the food plants. One can envisage two broad categories of metabolic expense involved in detoxication. First, there is the cost of synthesizing the appropriate detoxication enzymes and any associated morphological structures, added to the negative entropy of organization of the process; this is a "fixed cost" for any individual, determined by its genotype. Second, there is the metabolic cost of the detoxication process itself; this cost is variable, depending on the quantity of toxin metabolized by any particular phenotype and also, perhaps, on the energy content of the toxin itself. To the extent that the metabolic costs of detoxifying a given amount of a particular plant toxin remain constant from one insect to another, the variable costs of detoxication incurred by a specialist insect are unlikely to be much different from those encountered by a generalist insect feeding on a particular plant tissue. It is the underlying fixed costs which are more likely to represent a metabolic saving for the specialist since it need retain mechanisms for tolerating only a restricted range of potential toxins. The metabolic load carried by generalists might be reduced to some extent by a system for enzyme induction whereby detoxication enzymes are produced only when induced by the presence of an appropriate substrate (Yu and Terriere, 1973; Brattsten and Wilkinson, 1973); costs would thereby be transferred in part from the fixed to the variable category, imparting more phenotypic flexibility.

A survey of the activity of one component of the microsomal mixed function oxidase system in the larvae of 34 species of Lepidoptera has revealed considerably higher activity in the generalist species than in the specialists (Krieger, Feeny and Wilkinson, 1971) though the magnitude of the costs involved is not known. Even if such costs are very small, however, they may have a significant effect on fitness. In several (but not all) insect populations which have become resistant to insecticides, resistant insects have been found to lose their resistance rapidly when application of insecticides was discontinued (Brown, 1971), suggesting that maintenance of resistance may be associated with a cost in terms of population fitness.

There are likely to be fixed and variable costs, also, involved in the exploitation of the nutrient content of host plants. Increased variable costs, and resulting decrease in fitness of an individual feeding on any particular plant, will be related to how far the nutrient content and texture of the plant differ from the range of nutrient concentrations and plant texture to which that insect is optimally adapted. Once again, though, such variable costs may be similar for both generalists and specialists on any given acceptable plant. For example, J. M. Scriber (personal communication) has recently compared for a variety of Lepidoptera the efficiencies with which they convert digested food into biomass as a function of plant water content (Fig. 1). Though reduced water content evidently imposes additional metabolic costs on an insect it appears that such costs are of the same order of magnitude for generalists as they are for various more specialized species feeding on plants of similar water content (Fig. 1). Again, it is the fixed costs which may be appreciably higher for the general feeder but, as with the fixed costs associated with detoxication, we have little idea as to their magnitude.

At least in a qualitative way, therefore, a variety of possible adaptive advantages can be seen to accrue to insects with narrow host plant ranges. The persistence of many species of rather general feeding habits, however, prompts one to ask whether or not at least in some circumstances specialist feeders may remain specialists not because this is their optimal strategy but because once they have become specialists they have little evolutionary opportunity to reverse the process.

EVOLUTIONARY CONSTRAINTS ON SPECIALIST FEEDERS

The black swallowtail butterfly, *Papilio polyxenes* F., is an oligophagous species; its larvae attack only plants of one subfamily of the Umbelliferae. When larvae of this species were reared on celery plants (family Umbelliferae) which had been cultured in aqueous solutions of sinigrin (a thioglucoside restricted chiefly to the family Cruciferae and not known from umbellifers), feeding rates were scarcely reduced but the leaves became increasingly toxic as the concentration of sinigrin increased (Erickson and Feeny, 1974). At concentrations of sinigrin similar to those known from some crucifers, larval survival was negligible. This suggests that if larvae of *P. polyxenes* were to attack crucifers naturally, at best their fitness would be greatly reduced and at worst it would be zero, the outcome depending on the concentration of mustard oil glucosides.

Brower *et al.* (1967) were able to select for monarch butter-

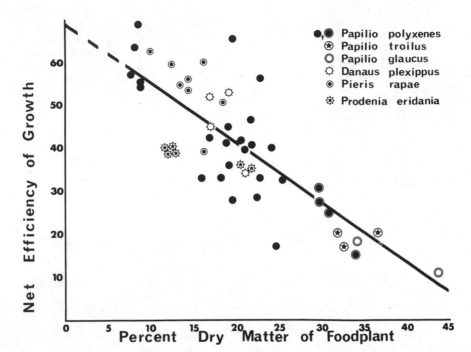

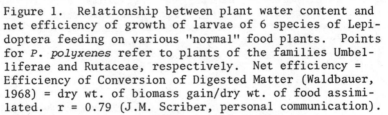

Figure 1. Relationship between plant water content and net efficiency of growth of larvae of 6 species of Lepidoptera feeding on various "normal" food plants. Points for *P. polyxenes* refer to plants of the families Umbelliferae and Rutaceae, respectively. Net efficiency = Efficiency of Conversion of Digested Matter (Waldbauer, 1968) = dry wt. of biomass gain/dry wt. of food assimilated. r = 0.79 (J.M. Scriber, personal communication).

fly larvae which could feed on cabbage, though enormous mortality occurred in the process; I suspect that a similar tolerance to sinigrin could be selected for in black swallowtail larvae and that if enough behavioral "mistakes" occurred in the field some larvae would, through mutation or recombination, be able to tolerate ambient concentrations of sinigrin with little reduction in fitness. However, this by itself would be unlikely to broaden the host range of the population, since it seems highly unlikely that resulting females would oviposit back on crucifers (or that the next generation of larvae would prefer these plants) and so any incipient broadening of host range would be genetically swamped. In other words, there seem to be at least two major kinds of evolutionary breakthrough involving secondary chemicals which must occur more or less simultaneously to enable a specialist insect to broaden its host range to include new plants of substantially different secondary chemistry. Individuals must first evolve the ability to tolerate ambient levels of toxins in the new plant without substantial loss of fitness and at the same time appropriate behavioral evolution (see Dethier, 1970) must occur to permit discovery of the new plant by the next generation of insects. It could be that the chances of both kinds of evolutionary breakthrough occurring simultaneously are remote. The chances might be very much higher were successional habitats to consist *only* of crucifers and umbellifers; not only would more frequent association with crucifers and their toxins occur, increasing the likelihood of a toxicological breakthrough (see Southwood, 1972), but discovery of crucifers by the next generation could follow simply from loss of obligate attraction to umbellifers. In practice, though, successional habitats usually contain many plant species which differ greatly in their secondary chemistry. For a specialist insect this may be a risky kind of environment in which to experiment with polyphagy.

Such experiments, however, are clearly more adaptive than certain death through starvation, and individual insects of many species are known to become more polyphagous when their normal host plants have become scarce or exhausted (Dethier, 1970). Under such circumstances, especially if the insect population is small enough so that new mutations or recombinations are unlikely to be swamped rapidly, one might expect permanent colonization of new host plants to be more likely even though initial fitness on such plants is low (Ehrlich and Raven, 1965).

It is possible, then, that even if it was adaptively advantageous for the black swallowtail to extend its host plant range to include crucifers such extension would be unlikely to occur because of lack of evolutionary opportunity (Erickson and Feeny, 1974). How far one can generalize from examples such as this must depend on the extent to which plants are potentially toxic to non-adapted insects. Several instances are known in which

oligophagous insects have been reared on "abnormal" host plants,
sometimes with little or no potential reduction in fitness (e.g.
Waldbauer, 1962). If it is generally true that toxicological
barriers to colonization of new plants are rare then the balance
between specialized and generalized feeding strategies may well
be determined by probability considerations such as those outlined by Levins and MacArthur (1969). My suspicion, however, is
that toxicological barriers in many plant communities are common
and that plant species in such communities are indeed "chemically
defended islands" (Janzen, 1968); in such chemically diverse communities one might expect biochemical coevolution to play the
major role in determining insect-plant relationships. Changes
in the host plant range of an insect would be most likely to
occur among plants of similar secondary chemistry to which evolution is catalyzed by partial preadaptation at the toxicological
and/or behavioral levels (see also Dethier, 1970).

EVOLUTION OF SPECIALIZATION

The evolutionary trend in host plant ranges of phytophagous
insects has probably been towards increasing specialization
(Dethier, 1954). The likely role of biochemical coevolution in
this process may be visualized in the form of a simple model
which is really an elaboration of the general scheme proposed by
Ehrlich and Raven (1965).

Consider an insect population which attacks many species of
plant. For a variety of reasons, such as variation in habitat,
nutrient content and plant texture, the fitness of the insects is
unlikely to be equal on all plant species, and we may expect
selection to favor some preferential discrimination of the plant,
say species A, on which fitness is greatest. Concentration of
attack might then be expected to have two kinds of effect on
plant A: it could cause A to become rarer (and thus harder to
find) and/or it might select for resistance, let us say chemical,
in A such that A is no longer the optimal host plant species
(Pimentel, 1961). Insects feeding on plant B now leave more
viable offspring than those on any other food plant, including A.
We may suppose that B now becomes rarer and/or evolves resistance
while A may become more common and its resistance might be re-
laxed. One can imagine this kind of dynamic equilibrium (Dethier,
1954) between the insect population and its various food plants
continuing indefinitely.

Meanwhile, it is likely that chemicals are being evolved by
the plants in response to many other selective pressures such as
competition with other plants or attack by various species of
bacteria and fungi or by other species of herbivore. Against any
such chemicals which are potentially toxic to it, however, our

insect population is able to evolve countermeasures because it is in frequent association with the plants as the compounds are evolved and elaborated. The insects, of course, may have to pay a metabolic price for their continued association with all these plants, and this could be reflected in reduced fitness on all food plants.

Suppose now that plant A's geographic range changes in relation to that of the insect population. Coevolution is now broken off and any chemical changes in plant A evolved for whatever reason in the absence of the insect population may pose a potential barrier to the insects if their ranges should once again overlap. If such changes had occurred only at the toxicological level or only at the behavioral level it is conceivable that, through partial preadaptation as discussed earlier, the insect population could adapt to them and bring plant A back into its food plant range. If, on the other hand, *both* kinds of change had occurred, plant A is likely to have been lost permanently from the feeding range. Clearly, though, if part of the insect population had been isolated along with plant A by some geographical barrier, for example, this population of insects would have coevolved with A in isolation and on recontact (assuming reproductive isolation between the insects) both species of insect might continue to coexist, one on A and the other on the remaining species of the original host plant range. As plant ranges expand or contract in relationship to those of insects, as geographical barriers come and go, as differential selection by the insects themselves as well as by other components of the communities bring about differentiation of secondary chemical content, one can easily see, I think, how specialization of feeding range can originate and even become intensified by positive feedback between plants and insects. It may not even be necessary to invoke geographic isolation for loss of a food plant species; if for whatever reason the fitness of an insect population remains low on a particular plant for a long enough period of time, ability to use this plant may be lost completely through effective lack of coevolution.

Though this model is greatly simplified and almost certainly overemphasizes the role of biochemical coevolution, it is consistent with several phenomena which might otherwise be puzzling. It suggests, for example, a reasonable explanation for the phenomenon of larval conditioning (sometimes referred to as the Hopkins Host Selection Principle) known to occur in several insect species (Dethier, 1954; Erickson, 1973). This process, whereby females prefer to lay their eggs on the plant species on which they fed as larvae, may have the effect of allowing a population to capitalize at any one time on those plant species within its overall host plant range upon which fitness is highest, while retaining contact and therefore coevolution with other

plant species in its host range. Such phenotypic flexibility
could be adaptive in a dynamic world in which what is the best
plant "today" might have evolved further resistance or become too
rare "tomorrow."

Secondly, one can very easily account for the apparent loss
of chemical defense by plants which are protected by other means.
Ant-acacias, for example, appear to have lost the chemical defenses characteristic of non-ant acacias (Rehr, Feeny and Janzen,
1973). *Dentaria*, a genus of woodland crucifers represented by
two species in central New York State, appears to escape substantial attack by the crucifer-feeding insect fauna both by its
habitat (which is atypical for crucifers) and by its early phenology (R. B. Root, personal communication). When *Dentaria* is
transplanted into open areas, however, it is rapidly decimated by
insects (Hicks and Tahvanainen, 1974). Moreover when larvae of
the European cabbage butterfly, *Pieris rapae* (which do not normally encounter *Dentaria* in the wild), are reared on *Dentaria*
leaves in the laboratory their growth rates and feeding efficiencies are higher than those on many of their usual crucifer food
plants (F. Slansky, unpublished results). These observations are
consistent with the idea that chemical defenses place a metabolic
burden on plants and that they are reduced or lost when the metabolic cost outweights adaptive benefit. Such an explanation has
been advanced, also, to account for the observed polymorphism
between cyanogenic and acyanogenic forms of certain legume species (Jones, 1972).

Thirdly, the coevolutionary scheme is consistent with a
recent observation by Janzen (1973) on the nature of interspecific competition between insects. Although many have suspected
that the adaptive radiation of phytophagous insects has in some
way been influenced by interspecific competition the puzzling
phenomenon remains that except in plant monocultures (where local
outbreaks may occur) or in plant communities which have been
modified by man, leaf-eating insects rarely seem to be sufficiently abundant to affect each other's food supply. Janzen (loc.
cit.) suggests that competition occurs, rather, in evolutionary
time in the sense that chemical defenses evolved by a plant as a
result of attack by one species of insect may automatically reduce the suitability of this plant for another insect species,
even though the two insect species may never occur on the same
plant tissues or even at the same time of year.

PLANT DEFENSE STRATEGIES

In attempting to decide to what extent the observed food
plant ranges of herbivorous insects are restricted by ecological
selective forces and to what extent they may be restricted by

evolutionary barriers it is important, I think, to consider the kinds of defenses against insects which may be employed by plants in different communities (see also Levin, 1971). On the basis of rather limited data, I would like briefly to contrast the kinds of defenses which might be expected in temperate zone early successional communities with those which seem to prevail in late successional and climax forest vegetation, also in temperate regions.

An early successional herb species is subject to attack both by insects which are adapted to tolerating its rather specific chemical defenses and by insects which are not so adapted. Against the first category of insect enemy the best strategy is that of escape; in ecological time this is achieved by rapid growth to maturity and production of large numbers of small seeds, many of which may be expected to escape, as seedlings, discovery by specialist insects. Ecological escape is aided, also, by association with other plants of different secondary chemistry, odors from which may mask discovery (Brues, 1920; Tahvanainen and Root, 1972). In evolutionary time, escape may be facilitated by a rate of evolution comparable to that of the plant's insect enemies, allowing the plants to be relatively sensitive to changes in selective pressures exerted by the insects. Against non-adapted insects, however, the best strategy (given the limited energy budget available) may be to retain relatively unique or unusual chemicals in quantities sufficient to repel a would-be attacker or, if attack persists, to interfere with insect growth and development. Since such non-adapted insects do not possess mechanisms for tolerating the plant's unusual chemical defense the quantities of chemical produced by the plant need not be large. This is essentially a "qualitative" defense which, if overcome by adaptation (as in insect species which have successfully colonized the plant in evolutionary time) is unlikely to affect greatly the "variable" costs incurred during insect growth.

Against non-adapted insects also it is advantageous for the ephemeral successional plant to associate with chemically very different plants. In such communities, where the strategic emphasis is on being "hard to find", the various plant species may be involved in a bizarre kind of mutualism. Although competing for light and nutrients, the more plant species there are and the more chemically diverse they can be, the harder it is for non-adapted insects to colonize new plant species and for adapted specialist insects to find their hosts. The further such positive feedback proceeds, the more firmly locked become the insects into rather rigid biochemical coevolution with their host plants, whether or not they might derive ecological advantages from colonizing new hosts if they had the evolutionary opportunity.

Most agricultural crop varieties have originated from successional herb species. It is interesting to speculate as to whether the vulnerability of crops to insect herbivores (in the absence of synthetic chemical defenses supplied by man) may result from planting in monoculture species which have evolved chemical defenses appropriate to communities in which the optimum strategy is being hard to find (Root, 1973; R. B. Root, personal communication). Moreover, such natural chemical defenses may have been further reduced over the centuries by man's selection for increased yield or improved flavor.

In contrast to many successional herbs, the dominant trees of a temperate zone climax forest are "bound to be found" by insects in ecological time, owing to their large size, long life relative to those of insects, and possibly also as a result of lower vegetational diversity. They are bound to be found in evolutionary time also and Southwood (1972) has shown that the number of insect species associated with the tree species of a temperate zone forest is roughly proportional to the abundance of each tree species in recent geological time. The optimum strategy for a dominant tree in such a forest, therefore, may be to make its leaves as inedible as possible as early as possible each season. This it appears to achieve by means of tough leaves, low in nutrients such as nitrogen and water, and by deploying relatively large quantities of chemicals such as tannins or resins (see Feeny, 1970; Southwood, 1972). Such compounds are generalized in their action and their effects are not simply countered by specific adaptation (Feeny, 1969). Insects attacking mature foliage of such plants must be adapted not only to any specific or "qualitative" chemical defenses which may be present but also to tolerating the high "variable" metabolic costs imposed on them by a diet low in nutrients and further fortified by "quantitative" chemical defense. Such generalized textural, nutritional and secondary chemical defense seems to be shared by numerous forest tree species (see Brown, Love and Handley, 1962; Southwood, 1972). To the extent that such general and shared defenses replace rather than supplement the more specific and diverse defenses of early successional plants biochemical coevolution might be expected to play a less rigid role in determining insect feeding strategies.

To portray the manner in which chemical defense may depend on community structure I have emphasized successional status and ignored relative abundance. Clearly, though, a rare (but persistent) climax species may be as "hard to find" as a commoner (though ephemeral) annual and might, therefore, be expected to deploy defenses more typical of the latter. Likewise the chemical defenses of a persistent or abundant successional species might be found to resemble those more typical of climax temperate zone species (see also Root, 1973).

This discussion has been greatly simplified and at times blatantly speculative. Central to its theme is the assumption that a majority of land plants contain secondary chemicals which serve as a defense against insect attack yet 14 years after Fraenkel (1959) gave this idea renewed prominence the available evidence for it remains meager and even ambivalent. I have assumed further that realistic predictions can be made solely on the basis of biochemical coevolution between plants and insects yet other herbivorous animals, fungi and pathogens must also play important roles in determining patterns of chemical defense by plants. Though direct evidence may be lacking, however, the circumstantial evidence now seems to point to a correlation between the persistence and relative abundance of a plant species and the kind of chemical defense it employs. This in turn could mean that the manifestation of biochemical coevolution between plants and insects may differ substantially from one kind of plant community to another.

SUMMARY

The role of biochemical coevolution as a determinant of the feeding strategies of herbivorous insects may depend on plant community persistence and diversity. Plant species which are rare or ephemeral or both are characterized as being "hard to find" by insect herbivores; chemical defenses of such plants are likely to be diverse and "qualitative", posing evolutionary barriers to non-adapted insects but only minimal ecological barriers to adapted species (against which the primary strategy is escape in time and space). Plant species which are abundant or persistent or both are, by contrast, "bound to be found" by insects both in ecological and evolutionary time; such plants appear to have evolved "quantitative" defenses, including tough leaves, low in nutrients and fortified with relatively large amounts of unspecific chemicals such as tannins. Such defenses pose a significant ecological barrier to insect herbivores though evolutionary barriers to colonization by insects are likely to be low unless supplemented by "qualitative" chemical defense.

ACKNOWLEDGEMENTS

I would like to acknowledge financial support from NSF Grant GB-33398 and to thank K. Arms, J. G. Franclemont, D. H. Janzen, P. Opler, R. B. Root, F. Slansky, Jr., and R. H. Whittaker for valuable discussion and insight.

LITERATURE CITED

Brattsten, L. B. and Wilkinson, C. F. 1973. Induction of microsomal enzymes in the southern armyworm (*Prodenia eridania*).
Pestic. Biochem. Physiol. (in press).

Brower, L. P. and J. V. Z. Brower. 1964. Birds, Butterflies,
and Plant Poisons: A study in ecological chemistry.
Zoologica 49:137-159.

Brower, L. P., Brower, J. V. Z., and Corvino, J. M. 1967. Plant
poisons in a terrestrial food chain. Proc. Natl. Acad. Sci.
U.S. 57:893-898.

Brown, A. W. A. 1971. Pest resistance to pesticides. *In* Pesticides in the Environment (R. White-Stevens, ed.), Vol. 1,
part II, Marcel Dekker Inc., N.Y., 629 pp.

Brown, B. R., Love, C. W. and Handley, W. R. C. 1962. Protein-
fixing constituents of plants: Part III. Rep. on Forest
Res. (1962):90-93, H.M.S.O., London.

Brues, C. T. 1920. The selection of food plants by insects,
with special reference to lepidopterous larvae. Amer. Natur.
54:313-332.

Dethier, V. G. 1947. Chemical Insect Attractants and Repellents.
McGraw Hill (Blakiston), New York.

Dethier, V. G. 1954. Evolution of feeding preferences in
phytophagous insects. Evolution 8:33-54.

Dethier, V. G. 1970. Chemical interactions between plants and
insects. *In* Chemical Ecology (E. Sondheimer and J. B.
Simeone, eds.), pp. 83-102, Academic Press, London and N.Y.,
336 pp.

Ehrlich, P. R. and Raven, P. H. 1965. Butterflies and plants:
A study in coevolution. Evolution 18:586-608.

Eisner, T. 1970. Chemical defense against predation in arthropods. *In* Chemical Ecology (E. Sondheimer and J. B. Simeone,
eds.), pp. 157-217, Academic Press, London and N.Y., 336 pp.

Erickson, J. M. 1973. Host plant selection by the eastern black
swallowtail butterfly, *Papilio polyxenes* F. (Lepidoptera:
Papilionidae). Ph.D. thesis, Cornell University, 158 pp.

Erickson, J. M. and Feeny, P. 1974. Sinigrin: A chemical bar-
    rier to larvae of the black swallowtail butterfly, *Papilio
    polyxenes*. Ecology 55: 103-111.

Feeny, P. P. 1969. Inhibitory effect of oak leaf tannins on the
    hydrolysis of proteins by trypsin. Phytochem. 8:2119-2126.

Feeny, P. 1970. Seasonal changes in oak leaf tannins and nutri-
    ents as a cause of spring feeding by winter moth cater-
    pillars. Ecology 51:565-581.

Fraenkel, G. S. 1959. The raison d'être of secondary plant sub-
    stances. Science 129:1466-1470.

Fraenkel, G. S. 1969. Evaluation of our thoughts on secondary
    plant substances. Ent. expl. et appl. 12:474-486.

Hairston, N. G., Smith, F. E., and Slobodkin, L. B. 1960.
    Community structure, population control, and competition.
    Amer. Natur. 94:421-425.

Hicks, K. P., and Tahvanainen, J. O. 1974. Niche differentiation
    by crucifer-feeding flea beetles (Coleoptera: Chrysomelidae).
    Amer. Midl. Natur. (in press).

Janzen, D. H. 1968. Host plants as islands in evolutionary and
    contemporary time. Amer. Natur. 102:592-595.

Janzen, D. H. 1973. Host plants as islands. II. Competition in
    evolutionary time. Amer. Natur. (in press).

Jones, D. A. 1972. Cyanogenic glycosides and their function.
    In Phytochemical Ecology (J. B. Harbone, ed.), pp. 103-124,
    Academic Press, London and N.Y., 272 pp.

Krieger, R. I., Feeny, P. P., and Wilkinson, C. F. 1971. Detoxi-
    cation enzymes in the guts of caterpillars: An evolutionary
    answer to plant defenses? Science 172:579-581.

Levin, D. A. 1971. Plant phenolics: An ecological perspective.
    Amer. Natur. 105:157-181.

Levins, R. and MacArthur, R. 1969. An hypothesis to explain the
    incidence of monophagy. Ecology 50:910-911.

Pimentel, D. 1961. Animal population regulation by the genetic
    feedback mechanism. Amer. Natur. 95:65-79.

Rehr, S. S., Feeny, P. P., and Janzen, D. H. 1973. Chemical defense in Central American non-ant acacias. J. Anim. Ecol. 42:405-416.

Root, R. B. 1973. Organization of a plant-arthropod association in simple and diverse habitats: the fauna of collards (*Brassica oleracea*). Ecol. Monogr. 43:95-124.

Rothschild, M. 1972a. Secondary plant substances and warning colouration in insects. *In* Insect/Plant Relationships (H. F. Van Emden, ed.), p. 59-83, Blackwell Scientific Publications, Oxford, 215 pp.

Rothschild, M. 1972b. Some observations on the relationship between plants, toxic insects and birds. *In* Phytochemical Ecology (J. B. Harbone, ed.), pp. 1-12, Academic Press, London and N.Y., 272 pp.

Schoonhoven, L. M. 1972. Secondary plant substances and insects. *In* Recent Advances in Phytochemistry (T. J. Mabry, ed.), 4, North-Holland, Amsterdam.

Southwood, T. R. E. 1972. The insect/plant relationship - an evolutionary perspective. *In* Insect/Plant Relationships (H. F. Van Emden, ed.), p. 3-20, Blackwell Scientific Publications, Oxford, 215 pp.

Stahl, E. 1888. Pflanzen und Schnecken. Biologische Studie über die Schutzmittel der Pflanzen gegen Schneckenrass. Jena. Z. Med. u. Naturw. 22:557-684.

Tahvanainen, J. O., and Root, R. B. 1972. The influence of vegetational diversity on the population ecology of a specialized herbivore, *Phyllotreta cruciferae* (Coleoptera: Chrysomelidae). Oecologia 10:321-346.

Verschaffelt, E. 1911. The cause determining the selection of food in some herbivorous insects. Proc. Acad. Sci. Amsterdam 13:536-542.

Waldbauer, G. P. 1962. The growth and reproduction of maxillectomized tobacco hornworms feeding on normally rejected non-solanaceous plants. Ent. exp. et appl. 5:147-158.

Waldbauer, G. P. 1968. The consumption and utilization of food by insects. Adv. Insect Physiol. 3:229-282.

Whittaker, R. H. and Feeny, P. P. 1971. Allelochemics: Chemical
    interactions between species. Science 171:757-770.

Yu, S. J. and Terriere, L. C. 1973. Phenobarbital induction of
    detoxifying enzymes in resistant and susceptible houseflies.
    Pestic. Biochem. Physiol. 3:141.

REMARKS ON CAROTENOIDS IN THE EVOLUTION OF SIGNALS

Miriam Rothschild

Ashton, Peterborough
United Kingdom

The wild asses did stand in high places,
they snuffed up the wind like dragons;
their eyes did fail because there was no grass
                    (Jeremiah 14.vi)

INTRODUCTION

As schoolgirls we were riveted by the idea that God's command "Let there be light" was a terse way of intimating that He
had thought up the eye. Later on my own children pushed the
concept back in time and suggested in their oblique fashion that
the question of light and darkness rested not on the creation
and/or evolution of the eye, but quite simply on that of
carotenoids, and the role of vitamin A in photosynthesis and
later in photo-reception.[1] I have always admitted candidly that
any mildly fruitful idea I have ever had, came from listening
attentively to the tiresome chatter of my children.
    Clearly in this case their imagination was fired because
everything was poetically yellow - the sun, the egg yolks, the
daffodils, globe flowers, the canary, the goldfish, cream,
cocoon silk, pollen, ripening corn and autumn leaves, and all
the other marvellous objects listed with so much enthusiasm by
Fox and Vevers (1960).
    I liked the idea because, for a student of warning colora-
tion, it suggested a good subsidiary question: What role do
carotenoids play, generally, in signals?

SIGNALS AND SIGNALLING

Plants and animals expend considerable time and energy in
attracting attention. Colors, scents, sounds, vibrations and
elaborate specialized movements are all geared for signalling of
one sort or another. Sometimes it is merely a matter of arousing
interest, but generally the objective is to get a specific
message across. The distance concerned may be a few centimeters,
or over two miles in the case of courting Gipsy moths (Kettlewell,
1973; Butler, 1970). Often the same sort of signal is used to
advertise both an attractive or unattractive quality - the sender
of the signal relying on an innate response or the experience of

the recipient to qualify or interpret the message. Thus Bilz says
(1963) "In connection with man's choice of food, not only colour
valencies but also danger valencies determine the reaction of the
subject". A good example is seen in the bright and varied colors
of both children's sweets and lethal pills (Rothschild, 1967; and
Plate 1, Fig. 1). In the first instance the manufacturer takes
advantage of youthful appreciation of bright colors and the
ability to learn to associate their brilliant hues and delicious
taste. In the second instance the chemist depends on the adults'
knowledge and experience to remind them by means of the eye-
catching reds, yellows and blues of their capsules, that seconal,
nembutal and sodium amytyl (The Chemist and Druggist, 1956, June
p. 604-605) are not to be trifled with. Similarly the berries
and fruits of some plants are colored red or black, or both, to
attract the bird disseminator of seeds (Snow, 1971), while other
scarlet or purple berries warn various species of their dangerous
and inedible qualities.[2] Thus the same signal can serve a dual
purpose, the message received depending on the recipient's life-
style and sensory equipment. In some cases the signals are confused, and children, for instance, may die from eating lethal
pills and - more rarely - inexperienced birds become dangerously
ill or perish from a surfeit of toxic insects.

     Signals, by and large, are a sophisticated side line of life,
which is carried to extremes of refinement when deceptive mes-
sages are promulgated as in the case of mimicry of color, sound,
form and movement. This observation applies equally to sexual
displays, which serve to bring male and female into close contact;
the less sophisticated animals just meet, or bump into each other.
One feature is common to all signals - they imply an "ulterior
motive". No plant or animal could afford to emit "pointless" or
random signals merely as an expression of exuberance; such a
tendency would be rapidly eliminated by natural selection. The
evolution of signals is something in the nature of an embellish-
ment, a luxury, a painting of the lily, for it is fraught with
dangerous side-effects. While signalling to a prospective mate
the individual may attract the unwelcome attentions of a predator. While advertising dangerous and repellent qualities the
warningly colored species exposes itself to many forms of
accidental death. For example, the aposematic eggs of the White
butterfly (*Pieris brassicae*) are frequently knocked off the surface of leaves by heavy rainstorms or hailstones (Baker, 1960), a
risk to which the cryptic and well concealed eggs of its relative
the Small White (*P. rapae*) are not exposed. Aposematic species
are victims of heavy predation by other insects. Witness the 33
species of Hymenopterous parasites known to attack the larvae and
pupae of the Large White, compared with the 18 species recorded
for the Small White (Table 1). Because so much signalling is, so

TABLE 1. Principal hymenoptera parasites of larvae of
P. *brassicae* and P. *rapae* (after W. R. Thompson, 1943-
1951; compiled by R. W. Crossky)

P. *brassicae*	P. *rapae*
Anilastus ebeninus Grav.	*Anilastus ebeninus* Grav.
Apanteles fulvipes Hal.	
Apanteles glomeratus L.	*Apanteles glomeratus* L.
Apanteles juncundus Marsh	
Apanteles rubripes Hal.	
Apanteles spurius Wesm.	*Apanteles spurius* Wesm.
	Apanteles rubecula Marsh
Bracon habetor Say	
Campoplex mutabillis Holmgren	
	Casinaria tenuiventris Grav.
Diadegma rufipes Grav.	
	Diadegma tenuipes Thoms.
Dibrachys boarmiae (Walk.)	*Dibrachys boarmiae* (Walk.)
Dibrachys cavus (Walk.)	
Eupteromalus germanicus Graham	
Exetastes illusor Grav.	
Exetastes nigripes Grav.	
Gelis agilis F.	
Gelis instabilis Foerster	
Gelis trux Foerster	
Gelis areator Panz.	*Gelis areator* Panz.
Habrocytus dispar Curtis	
Habrocytus vibulenus (Walk.)	
	Habrocytus chrysos Walk.
Hemiteles cingulator Grav.	
Hemiteles imbecillus Grav.	
Hemiteles melanarius Grav.	*Hemiteles melanarius* Grav.
Hemiteles pictipes Grav.	
Hemiteles socialis Ratz.	
Hemiteles tristator Grav.	
	Hemiteles schaffneri Schmied.
Lysibia nana Grav.	*Lysibia nana* Grav.
Mesochorus tachypus Holm.	*Mesochorus tachypus* Holm.
Mesochorus anomalus Holm.	
Mesochorus curvulus Thoms.	
Mesochorus tuberculiger Thoms.	
	Mesochorus pectoralis Ratz.
	Mesochorus splendidulus Grav.
Stictopisthus complanatus Hal.	*Stictopisthus complanatus* Hal.
	Tetrastichus rapo Walk.
Theronia atalantae Poda	*Theronia atalantae* Poda

to speak, a refinement grafted on to existing patterns of behavior, it is sometimes possible to speculate correctly on the course
of their evolution, which makes the subject attractive to the
naturalist. One example will suffice to illustrate this point:
when a pigeon rises vertically off its perch or nesting site,
scared out of its wits by the presence of some raptorial bird, its
wings meet above its head with such force that an audible clap
results. Weis Fogh (1974) has recently demonstrated the aerodynamics of flapping flight and shown how the meeting of the wings
in this fashion improves performance. In fact this important
feature of flapping flight is common to butterflies[3] and various
other insects, as well as birds. The wood pigeon has, most
effectively, tacked on the clap to its sexual display. Those of
us who have the good fortune to live in the country in the UK,
have watched this delightful and familiar bird[4] towering in
leisurely fashion, clapping and gliding to show off to its female
companion. This is a simple case of a sexual display evolving
via a successful escape mechanism. It is not usually so easy to
unravel the course of events. All the text books, for instance,
tell us that the scarlet flowers which attract bird pollinators
are scentless: in a somewhat reminiscent situation we find that
the stickleback (*Gasterosteus aculeatus*), of which the male
attracts the female by displaying its scarlet ventral surface,
virtually lacks the sense of smell (Bardach and Todd, 1970).
Which part of the pattern evolved first? Had the scent element in
these signals been lost because color signals proved more effective, and selection consequently no longer found it necessary to
sharpen the quality of odor? Or was color operating from the
first and the development of a keen sense of smell and smelling
therefore unnecessary? It is true that, living in shallow water,
where both fish and molluscs can take advantage of the greater
light intensity, these animals frequently use colors as a means of
signalling, but we are still left wondering.

  There is another type of dual purpose "signalling" which
merits more careful consideration. It is much safer for an animal
to remain unseen while a predator is at a safe distance, even if
at close quarters aposematic coloring and warning displays prove
advantageous. Thus an organism can, by another method, use much
the same signals for a totally different purpose. In the example
cited above, bright colors were used to arouse attention, and the
recipient was stimulated to take or reject the object offered for
appraisal. In this case the same bright colors serve to conceal
the object altogether from a distance, a "negative signal" for
want of a better word, and to repel the observer at close
quarters. A good example is the caterpillar of the cinnabar moth
(*Tyria jacobaeae*) which from a distance tones in marvellously with
the yellow flowers of the host plant, and is unquestionably cryp-

tic, but at close quarters the yellow and black stripes are wasp-like and positively aposematic. In the Alps one can appreciate a different expression of the same sort of situation. Here, seen at close quarters, walking about composedly on the snow, parties of Mountain Choughs with their black plumage, bright yellow bills and red legs, are startlingly conspicuous and unquestionably aposematic. Yet once in the air, viewed against alpine rock faces, piebald with intermittent patches and streaks of snow and ice, these birds become virtually invisible, displaying the very acme of cryptic concealment (see also Kettlewell, 1973, on aerial crypsis). In order to appreciate the role of dual purpose signalling it is necessary to consider the different types of eyes found among potential predators, and even the small anatomical and morphological differences between broadly similar eyes. As Daw (1973) says, tantalizingly little is known about color vision in birds, but it is necessary to treat each species individually. For a good field entomologist one of the most frustrating aspects of ageing is that the cryptic moth or caterpillar which we spotted so easily when we were young collectors, now "gets away with it". Almost as aggravating is the fact that unless we put the forceps and the watch glasses and cover slips in exactly the same place on the table we cannot find them. A little loss in accommodation has this disconcerting effect. This gives one some insight into the success of crypsis. It is difficult for us to appreciate how lost we could be without our critical faculties, by the aid of which we can, to some extent "improve" our eyesight or compensate for the loss or lack of accommodation (Gregory, 1972). Even so, the inability to sweep a glance round the table and immediately pick out the desired object is a great handicap. Furthermore, we cannot imagine how animals like birds, which are dependent solely on their acuity of vision, inborn perceptual skills and brief experience of a harsh world, interpret what they see, lacking our own powers of deduction and reasoning. Without these intellectual endowments, which give us information about size, space, distance and perspective, the "face" of the eyed hawkmoth in display, for instance, would "look" very different and much more frightening. Conversely we must try and imagine how important tiny details would become if our eyes were four times superior to what they were in our youth. It came as a great surprise to me, for instance, to realise that a bird can see which way a flea jumps (Lane, 1963) since this insect develops an acceleration of 140G when it takes off (Rothschild et al., 1973). Therefore *Copsychus malabaricus* would see one of our standard Hollywood movies as a series of clear, isolated pictures.

Because of the difficulty of appreciating the difference between what a bird can "see" and "perceive" we underestimate the

importance of dual purpose signalling, especially in the evolution of aposematic coloration. In the great majority of cases it
must pay the warningly colored species to remain hidden until the
predator is close at hand. Out of range and unseen no accident
can occur, but once the enemy is within striking distance it may
be advantageous to change quickly, and advertise toxic or
dangerous qualities before it comes too close. This can be
achieved by the rattle of the rattlesnake as well as the sudden
display of orange hind wings by a poisonous grasshopper. In fact
there must be a very critical distance at which signals are
switched from one sort to another.

It has been suggested elsewhere (Rothschild, 1971) that a
species ingesting and storing toxic secondary plant substances is
destined to adopt the aposematic life-style. This is because any
mutation which assists the predator in identifying and avoiding a
dangerous or distasteful species must have survival value. But
because of the peculiar relationship of crypsis-at-a-distance and
aposematic-close-up signals, a mutation producing only a tiny red
or yellow spot or yellow tinted haemolymph can prove advantageous.
All plants and animals are, in a sense, preadapted to this type of
response because carotenoids are always available on such a lavish
scale.

SIGNALS FROM PLANTS

In their classical paper on the evolution of butterflies and
plants Ehrlich and Raven (1965) emphasize the stepwise evolution
of plants and insects in their struggle for survival. In this
article the authors pay special attention to the synthesis of the
so-called "secondary plant substances" that protect them from
attacks by herbivores in general, and insects in particular.
They point out that the insects, for their part, were successfully countering these measures, and in a number of cases even
using the chemicals in question for their own advantage
(Reichstein, 1967). Where carotenoids are concerned, an even more
elaborate set of complicated symbiotic-like interactions can be
traced between insects, birds and plants, despite our ignorance
concerning their precise role in the bodies of many animals. It
must be recalled that until Fraenkel (1959) suggested a defensive
function for the secondary plant substances, these were generally
regarded as waste products of metabolism. Strangely enough, as
late as 1973, Dadd in his review of insect nutrition was almost
apologetic in offering his conclusion that the role of carotene
in insects "is physiologically more complex than an adventitious
food-derived waste".

One non-controversial function of carotenoids concerns the
major signalling device of many flowering plants, their colored

inflorescence and fruits (Table 2), by which they attract insect
pollinators and bird seed-disseminators. Plants, via carotenoids
and vitamin A, give both these groups of animals vision (Dadd,
1973) and then by means of yellow flowers, yellow pollen and red
and orange fruit, use carotenoids to signal back to them. It must
always be remembered, extraordinary though it may seem, that
animals cannot synthesize carotenoids and are entirely dependent,
either directly or indirectly, on plants for their supply. They
have, however, evolved both an elaborate and versatile genetic
control of these pigments.

Plants, from their foliage, have to provide caterpillars
with their carotenoids, without which the future butterfly or moth
could not receive the signals transmitted, and upon which they
depend for pollination. Rarely do caterpillars eat flower petals[5]
unless the rest of the plant has been entirely consumed, although
there are exceptions: the final instar of the Oleander Hawk moth
and Mullein moth do so, taking on the pink and yellow, respectively, as part of their ambivalent coloration, sometimes cryptic, sometimes warning. Broadly speaking petals have different
carotenoids than the foliage (Goodwin, 1965, 1966) and always less
than the corresponding leaves, and this is probably why they have
insufficient flavor to whet the caterpillar's appetite, for in
leaves carotenoids can function as feeding stimulants and, for
example, the moth *Plodia interpunctella* (Morère, 1971) fails to
feed on artificial diets lacking carotenoids, and dies of starvation. Dixon (personal communication) has suggested that the
petals probably lack other essential food ingredients, in which
case they would afford another good example of dual purpose
signalling - negative for caterpillars, but positive for adult
butterflies, although presumably "putting the message across" to
each stage via a different sensory apparatus.

The consideration of fruit and flowers must lead inevitably
to an appraisal of the plants' other obvious signalling devices -
scent and taste.

I have speculated elsewhere[6] that carotenoids may play an
important role in scent reception and scent enhancement
(Rothschild, 1971, p. 214-215; Rothschild *et al.*, 1972) and,
although there is as yet no hard evidence to support this rather
wild hypothesis, nothing has come up which suggests it is complete
nonsense. The protein-bound carotenoids in the nasal cavity of
both mammals (Moncrieff, 1951; von Skramlik, 1963) and many birds
(Bang, 1971) have been considered as possible receptors for
olfactant molecules inhaled via the nose. Briggs and Duncan
(1961) point out that 50 out of 56 patients recovered completely
or partially from uncomplicated anosmia when injected with
carotenoids, suggesting that olfaction might be included together
with photosynthesis and vision, as "yet another process in which

TABLE 2. Examples of carotenoids in plants (selected from Goodwin, 1950, 1952; Fox and Vevers, 1960; Weedon, 1971;
Valadon and Mummery, 1968)

Species	Location	Color	Carotenoids (selections only)
Epicoccum nigrum Fungus	general body tissues	pink	Rhodoxanthin
Torula rubra Red Yeast	general body tissues	red	Torularhodin
Chromatium warmingii Bacteria	general body tissues	purple	Rhodopinol Rhodopinal and other Xanthophylls
Spirobacillus cienkowskii Bacterium	body tissues of host	bright red	Rhodoviolascin
Capsicum annuum Red Peppers	fruit	red	Capsanthin Capsorubin Cryptocapsin
Buxus sempervirens Box	leaves	purple	Dianhydroesch- scholtzxanthin
Jacaranda ovalifolia Jacaranda	flowers	sky blue	β-carotene
Rosa rubiginosa Rose	berries (hard)	red	Rubixanthin
Taxus baccata Yew	berries (soft)	red	Rhodoxanthin
Narcissus majalis Pheasant's Eye narcissus	fringe of corona	red	β-carotene
Trollius europaeus Globe flower	petals	yellow	Trollixanthin (= neoxanthin?)
Ipomoea batatus edulis Sweet potatoes	tubers	yellow orange	β-carotene
Lilium tigrinum Tiger Lily	anthers	yellow	Antheraxanthin Capsanthin
Tulipa gesneriana Tulip	pollen	yellow black?	β-carotene Lutein
Brassica oleracea var capitata Cabbage	outer leaves	green	β-carotene Lutein (15 others have been identified)

the electron mobility that characterizes carotenoid molecules may
be utilized in a process of energy reception and transfer".
Rosenberg et al. elaborated the theory (1968) and there are
numerous scattered references which add a little circumstantial
and anecdotal evidence (McCartney, 1968). It is a delightful
idea that plants may have used the carotenoids for stimulating
insect sensory receptors, apart from the question of vision and
photo reception, and that animals can smell as well as see by
virture of yellow pigments.

When searching for nectar some insects respond to color, and
others to odor signals from plants (Proctor and Yeo, 1973). In
certain instances the scent of flowers operates from a distance,
but once attracted to the immediate vicinity, it is the colors
which induce the insect to settle, and stimulate it to probe.
Conversely, in other cases, a large patch of color attracts their
attention from far off (Parmenter, 1958): only at close quarters
does smell then become effective. It is interesting that various
flowers which grow in eye-catching stands, are relatively scentless, and some of their pollinators, at any rate, do not respond
to smell. A preference for particular colors is innate in certain
species of insects, and so is the attraction of dark spots on
flower petals to bees. It would be interesting to compare the
concentration of carotenoids in various parts of those flowers
which are both heavily scented and colored yellow by these pigments. It may well be found that, by and large, flavonoids color
the less heavily scented flowers as well as providing suitable
material for nectar guides. All insect pollinated blossoms, however, contain carotenoids in their pollen grains, and these substances are only lacking in cases in which the plants are
primarily wind and water pollinated (Barbier, 1970; Goodwin, 1952)
such as Pinus, Populus, Betula and Alnus. Pollen of sallows and
willows which are both insect and wind pollinated also contain
carotenoids. What is the function of the yellow pigments in
pollen grains? What meaning underlies this particular signal? In
the animal kingdom these substances are frequently concentrated in
the reproductive organs (Goodwin, 1950, 1952; Table 3) and therefore even if their function in these particular sites has not been
explained, it can come as no surprise to find carotenoids concentrated in the reproductive organs of plants. Their color attracts
attention to this protein-rich food source (Lepage and Boch, 1968)
which serves to fatten up and protect future pollinators - for
bees pass carotenoids into honey propolis and beeswax. But do
these pigments also boost the smell and taste of the trienoic
acid (Hopkins et al., 1969) present in pollen, and which is
thought to function as a food marker for bees?

Hinton (1973a and b) has suggested that the bright colors of
the inflorescence, as opposed to the dark lines and spots of the

TABLE 3. Examples of carotenoids found in some external and internal tissues (selected from Fox and Vevers, 1960; Goodwin, 1950, 1952; and Weedon, 1971)

EXTERNAL				INTERNAL			
Species	Organ or Tissue	Color	Carotenoids (selected)	Species	Organ or Tissue	Color	Carotenoids (selected)
Guara rubra Scarlet Ibis	plumage	crimson	Canthaxanthin Phoenicopterone Astaxanthin	*Gallus domesticus* Gintro penguin etc. etc.	yolk of eggs	yellow bright red	Lutein Zeaxanthin
Oriolus auratus Golden Oriole	plumage	golden-yellow	Lutein (crystalline)	Various birds and turtles	oil droplet filters in eye	yellow, orange and red	Sarcinene Lutein Zeaxanthin Astaxanthin
Phasianus colchicus Pheasant	wattles	red	Astaxanthin	*Gallus domesticus*	retina	yellow	Galloxanthin
Larus ridibundus Gull	beak, legs claws	yellow and red	Astaxanthin	Various birds and mammals	nasal cleft tissue	yellow to dark brown	β-carotene
Cissa chinensis Hunting Crow	plumage	green in forest area blue in open country	Carotenoids + blue structural color	*Fulmarus glacialis* Fulmar petrel	oily fluid in proventriculus	amber	Carotenes
Xipholena lamellipennis Chatterer (Cotingidae)	plumage	very dark red (almost black)	Carotenoids	*Homo sapiens* man	colostrum milk	straw white to yellowish	β-carotene α and β-carotene Lycopene Lutein
Crossaster pappposus Star Fish	dorsal skin	red and violet red	Carotenoids Chromoproteins		plasma and lymph	straw	Carotenes Xanthophylls
Hypsipops rubicunda Marine goldfish	skin	bright orange (adult specimen)	Xanthophylls Taraxanthin		corpus luteun	yellow	β-carotene
Sebastes norvegicus Scorpaenid fish	skin	bright red	Astaxanthin	Cattle	ear wax	yellow	Carotenes
Homarus gammarus Lobster	carapace antennae	blue bright red	Crustacyanin Astaxanthin	*Salmo salar* Salmon	muscle	pink to red	Astaxanthin

Organism	Site	Colour	Pigment
Pandalus borealis Shrimp	external carapace	bright red	Astaxanthin ?
Macrobrachium rosenbergi Giant Indian Prawn	external carapace	bright blue	Carotenoprotein
Hippolyte varians Prawn	external carapace	bright green	Carotenoid (yellow) Carotenoprotein (blue)
Lepas sp. Goose Barnacle	eggs	blue pink	Carotenoprotein
Oedipoda spp. Locusts	hind wings	red yellow blue	Astaxanthin Carotenoprotein
Andara broughtoni Gasteropod mollusc	"foot"	red/orange	Pectenoxanthin
Asterias rubens Starfish	integument	red	Astaxanthin
Actinia equina Sea anemone	general body color	red	Actinioerythrin
Hymeniacidon sanguineum Sponge	general body color	bright red	α-carotene γ-carotene
Pieris brassicae Large White Butterfly	eggs	bright yellow	Lutein β-carotene
Schistocerca gregaria Locust	eggs	yellow	β-carotene (when first laid) Astoxanthin (during development)
Bombyx mori (Silkworm)	eggs	yellow	Carotenes Xanthophylls
Metatranychus ulmi Apple tree mite	eggs	orange & bright red	β-carotene and others
Homarus gammarus Lobster	stored eggs	bright green	Ovoverdin (= Astaxanthin + Lipoprotein)
Aposematic insects (various)	haemolymph	bright yellow	Carotenoids (various)
Apis mellifera Honeybee	bees wax honey	yellow yellow to gold	Carotenoids β-carotene Lutein
Arctia caja Garden Tiger Moth	gonads	bright pink	Carotenoids (?)
Pieris brassicae Large White Butterfly larva	gonads of male	purple	Carotenoids (?)
Pentatomid and Coreoid Bugs (various)	lining of scent gland reservoir	bright orange yellow	Carotenoids (?)
Patella depressa Limpet	gonads ♀ gonads ♂	brown/green pink	β-carotene Zeaxanthin Echinenone Cryptoxanthin
Pecten maximus Giant scallop	gonads	bright orange/red	β-carotene Pectenoxanthin (= Alloxanthin?)
Pomacea canaliculata Freshwater prosobranch snail	stored & laid eggs	bright red	Ovorubin
Parasitic infusorians (*due to eating eyes of crustacea)	whole organism	red*	Carotenoids
Doris tuberculata Nudibranch molluscs	liver	bright green	Carotenoids
Cucumaria lactea Holothurian	gonads	bright blue	Carotenoids

nectar guides, were evolved in response to attack by herbivorous
reptiles, and constitute a warning mechanism which operated in
the dim and distant Cretaceous past. These animals, like birds,
have well developed color vision (Walls, 1963; Daw, 1973;
Hodgkin, 1971; Baylor and Hodgkin, 1973) and the writer can vouch
for the predilection of at least one species for carotenoids, for
the children's tortoises will slowly and systematically denude the
garden of yellow flowers, including the bitter tasting buttercups
growing on the lawns. Certainly seed dispersal by reptiles
(saurochory) must have preceeded that by birds. Today there are
still some good examples of this, and the saurochorous fruits
which grow low down on the trunk or stem, have attractive odors
and bright colors (van der Pijl, 1966) contrasting with the dull
tones, foetid odor and grandiose proportions of those fruits
adapted for seed dispersal by mammals. Even if the reptilian eye
had played a part in initiating color signals from plants (which
were equivalent to the warning odors of toxic foliage directed
towards mammalian herbivores today) what we see now must be the
result of powerful selection pressures exerted by the stepwise
evolution of the contemporary pollinators and the plants themselves. Numerous examples of balanced color polymorphisms among
present day species support this view. It is the dual purpose of
such signalling which gives rise to confusion in the minds of some
naturalists, for both distance and close-up signals are simultaneously involved. The dark lines and spots and other specialized
areas on the petals, reflecting or absorbing ultra violet light,
guide insects (which can see in the ultra violet) to the exact
location of the nectaries, while the distance signals are concerned with the concentration of brightly colored blooms in space
and time, which serve to attract them from afar.

The signals from plants are not, of course, confined to those
sent out by their flowers, although those emanating from leaves or
roots are almost certainly more generally designed to repel than
to attract. There are a fair number of exceptions such as the
scarlet leaves round the flowers of _Poinsettia_ which have ursurped
the function of petals. Most of these signals from the foliage
take the form of powerful odors and bitter taste, and are often
associated with toxic secondary plant substances (Table 4), and
plants rejected by many large herbivores (Rothschild, 1964, 1972a).
As we have suggested, these are probably directed towards
mammals rather than insects. This year we have chosen several
toxic plants with odoriferous foliage and compared their
carotenoid content with that of related plants with less smelly
leaves. We expected a high concentration to boost their warning
properties. The only species, however, which came up to our
expectations were the stinking Aristolochias, in which relatively
large concentrations of carotenoids were found (Table 4; Boothroyd

Table 4. Carotenoids in toxic and non toxic plants and aposematic insects (selected examples). Adapted from papers by Boothroyd and Swain (in prep.), Mummery, Rothschild, and Valadon (in prep.), Feltwell and Rothschild (1974), Marsh and Rothschild (1972b). The latter paper contains the references to the insect toxins and by whom tested.

Name of plant	Some toxins recorded from plants	Odor of foliage of plants	Total carotenoids in plants µg/g	Some aposematic insects feeding on plants	Toxins stored by insects	Total carotenoids in insects µg/g
Aristolochia clematitis (3 trials)	aristolochic acids present Magnoflorin	very strong foetid smell	442-492	*Zerynthia polyxena*	aristolochic acids	not known
Aristolochia elegans (3 trials)	aristolochic acids present in relatively small concentrations	less powerful, but strong	174-320	*Troides aeacus*	negative for aristolochic acid and magnoflorin	not known
Calotropis procera (3 trials)	various cardenolides present	very strong	128-140	*Poekilocerus bufonius*	calactin and calotropin	not known
Vincetoxicum nigrum (1 trial)	various cardenolides present	slight, foetid	297		none tested	not known
Hoya imperalis (3 trials)	lacking cardenolides	strong to very faint	77-89.3		none tested	not known
Stephanotis floribunda (2 trials)	lacking cardenolides	well defined but not powerful	119-187		none tested	not known
Asclepias curassavica (4 trials)	various cardenolides present	very variable, characteristic of family	182-465	*Danaus plexippus* ♂ *Aphis nerii* *Euchaetias antica*	cardiac glycosides " " " "	122.53 not known not known
Asclepias incarnata (2 trials)	lacking cardenolides when tested*	strong: characteristic of family	172-195	*Labidomera clavicollis* (Coleoptera)	none present in those tested*	not known

Plant (number of trials)	Chemical constituents	Odor		Insects feeding	Insect toxins	
Nerium oleander (3 trials)	various cardenolides present	ranging from faint to strong	71.1-95.2	Syntomeida epilais Caenocoris nerii Aphis nerii Aspidiotus nerii	cardenolides cardenolides cardenolides cardenolides	not known not known not known not known
Senecio vulgaris (1 trial)	pyrrolizidine alkaloids	faint odor	116	Tyria jacobaeae	pyrrolizidine alkaloids	427.30 (reared on S. jacobaea)
Passiflora foetida (3 trials)	?	powerful: foetid	191-225	Heliconius erato	nature of lethal factors not known	not known
Passiflora caerulea (3 trials)	toxic alkaloids not identified	less powerful but still strong	137-202	Heliconius erato	nature of lethal factors not known	not known
Urtica urens (1 trial)	5HT, ACh histamine	strong odor when wet	120	Aglais urtica	not tested	774.62 (reared on U. diocea not U. urens)
Brassica oleracea var capitata (large number of trials) Feltwell 1973	sinigrin; mustard glycosides	slightly disagreeable	250-6389	Pieris brassicae (all stages aposematic) Pieris rapae Pieris napi	sinigrin; mustard oils; unidentified toxin in pupa in pupa	179-245 66 24
Brassica oleracea var. Greyhound (plant not tested for carotenoids)	sinigrin; mustard glycosides	slightly disagreeable	not tested insects fed on parts of same leaves	P. brassicae pupa (100) P. napi pupa (25) P. napi pupa (25) (diapausing strain)	not tested not tested not tested	332-647 41-50 88
Tropaeolum majus (1 trial)	sinigrin	powerful leafy smell	6756	Pieris brassicae adult larva (4th instar) pupa (4 days after pupating)	not tested	871 (per insect 19) 1055 (per insect 25) 679 (per insect 27)

* Rothschild, Reichstein and von Euw, in prep.

The concentration of carotenoids varies so greatly in the same species of plants, and in the insects feeding on them, that no conclusions can be drawn from such small samples. Furthermore the insects should be reared on parts of the actual plants to be analyzed (not on different plants) and then themselves tested. Odors can only be assessed subjectively and are also very variable.

and Swain, in preparation). It will be noted that both species of
the Asclepiad plants lacking cardiac glycosides had the lowest
carotenoid content of the family so far tested. On the other hand
the same character is evident for the related Oleander, which is
very rich in cardenolides (Jäger *et al.*, 1959). This is virtually
an untouched field which probably merits more attention.

    Some of the specializations which are associated with plant
signals are very elaborate. Fruits which attract birds are
cryptically colored before they are ripe, and often extremely
sour or bitter at this stage. It is only when the seeds are
ready for dispersal via the intestine of the birds that they
ripen - often associated with a massive synthesis of carotenoids[7] -
and produce their brilliant reds and yellows and blackish blues
which attract the frugivorous species. Simultaneously the concentration of sugars increase in the pulp and the subtle fruity
scents are produced. Flowers which depend on bird pollination are
also highly specialized. One of the more impressive of these
mutual evolutionary patterns is the marked prevalence of fiery
red flowers among those pollinated by *migrant* humming birds. This
is another instance of signals which have to function from a
distance rather than at close quarters and which necessitate quick
recognition. Birds which pass rapidly from one area to another
benefit by a signal common to their food source (Porsch, 1931;
Grant and Grant, 1968) and also one which can be seen from afar.
It comes as no surprise to find that among the more static species
of humming birds, in tropical environments, color preference for
the carotenoid colored flowers is less pronounced. It should be
noted that humming bird flowers also lack the nectar guides
visible in the ultra violet, and found in many bee pollinated
flowers, and, as we have said these blooms are also scentless.

SIGNALS FROM ANIMALS

    Both birds and insects feeding on plants and other animals
make use of carotenoids for their own signals: these substances
provide most abundant coloring matter despite the strange fact to
which we have already drawn attention, that animals cannot
synthesize carotenoids. In some birds, however, the dietary pigments are transformed before they are deposited in the plumage.
The red, yellow and orange feathers of Goldfinches, the brilliant
plumes of the Cock-of-the-Rock (*Rupicola rupicola*) (Weedon, 1971),
the scarlet coloring of flamingoes and, surprisingly, perhaps, the
reds, violets and blues of Fruit pigeons, are good examples of the
signalling repertoire that owe their intense arousal qualities to
carotenoids. Sexual excitement and aggression are engendered by
the sight of these pigments, and they are also used for both
temporary and permanent recognition signals. Thus the male black-

bird's bill is yellow throughout the year, but both sexes of the
starling develop yellow beaks in the spring, linked to a rise in
blood androgens (Witschi, 1961). The nuptial display plumage of
the male Bishop Bird contains concentrated carotenoids which are
later withdrawn and stored in the liver and fat when the mating
season is over (Goodwin, 1952).

The male locust turns bright yellow when it becomes sexually
mature (Loher, 1960), an event which is linked to the production
of volatile pheromone, quite distinct from β-carotene, which
accelerates maturation in other specimens in the assemblage. The
coloration of the male Bullheads (*Ictalurus natalis*) changes to a
bright yellow in the presence of a pregnant or gravid female, and
he approaches her with evidence of sexual excitement which includes fanning and gasping movements (Bardach and Todd, 1970).

Carotenoids are also involved in communal recognition, flock
synchronization and feeding cues between parent and young. The
best known examples of the latter are the yellow gape of nestlings
and the red blotch on the beak of parent herring gulls (Tinbergen,
1953). Carotenoids, xanthophyls and traces of β-carotene are also
present in the oesophageal fluid with which flamingoes feed their
young, but its function in this secretion is unknown.

Thus in the vertebrates the carotenoids play their part in
the arousal of lust and rage and the gentler emotions of parental
care and solicitude.

Sometimes carotenoids seem to stimulate aberrant or, at any
rate to us, inexplicable behavior in birds. We have all witnessed
time and again the apparently pointless destruction of yellow
crocus flowers[8] by sparrows, while the purple and white varieties
remain unscathed. I have opened the crop of a wood pigeon and
found it crammed with flower heads of buttercups in full bloom -
as if the bird were "hooked" on the combination of carotenoids and
starch granules found in their petals. Cattle, of course, leave
these bitter tasting blooms severely alone, suggesting that even
if cows are color blind, the reflections from the petals give them
an easily recognizable cue or identification mark.

Among the invertebrates carotenoids are involved in aposematic color schemes more often than in either plants or vertebrates.
Thus poisonous sponges, toxic crinoids and sea cucumbers and
certain molluscs are wholly or partially brilliant yellow, red or
orange, and the excessively poisonous eggs of the moth *Zygaena
lonicerae* (Marsh and Rothschild, 1974) are also bright yellow in
hue.

Carotenoids are present in the haemolymph of all insects so
far examined (Feltwell and Rothschild, 1974; Goodwin,
1971b) but it is a striking fact that it is usually pale green,
straw colored or virtually colorless, but in toxic and dangerous
species it is frequently bright yellow. Good examples are the

"blood" of ladybirds and Arctiid moths (Rothschild, 1961, 1971, Plate 2, Fig. 3; Pasteels *et al.*, 1973; Valadon and Mummery, 1973).

Brilliantly colored internal tissues are of special interest and significance in the absence of an appreciative eye, since their pigmentation suggests important physiological attributes, fortuitously linked to, or associated with, vivid hues. The best examples are the red color of vertebrate blood[9] and the brownish tinge of insect muscle, in which cytochrome is concentrated. But what is the significance of the carotenoids coloring the scarlet gonads of the scallop (*Pecten*), the pink muscles of salmon, the yellow mammalian corpus luteum - the nasal cleft of mammals and birds - the purple and orange testis of many insects, or bright orange lining of their scent glands? We do not know. Much the same can be said for the yellow haemolymph (Table 5) of aposematic insects. But in this case part of its effect is exerted external-ly, where it can be seen, for it is oozed or squirted onto the surface of the body through specialized bleeding areas and gland apertures. There it often mixes with toxic and odoriferous defensive fluids (Rothschild, 1961, 1971), forming bright yellow conspicuous droplets (Plate 1, Fig. 3). In other cases the toxins are circulating in the haemolymph itself (Reichstein *et al.*, 1968; Jones *et al.*, 1962). The very keen eyesight of birds, combined with their specialized feeding habits (Swynnerton, 1915; Roths-child, 1971) is closely linked with this particular adaptation, for their attention is attracted by this method of, so to speak, underlining the aperture of toxic glands (Plate 1, Fig. 4). In everyday human life a similar device is used by the makers of cameras and other scientific apparatus. Small circular red or yellow or white spots of paint are placed near important holes where the machines in question must be oiled, or parts aligned (Plate 1, Fig. 2). Some insects, mimics as well as models, even evolve spots which simulate or reinforce the drops of yellow haemolymph or mark the apertures from whence it is expelled (Portchinsky, 1897).

It is extremely unlikely that this is the sole function of the carotenoids in the yellow haemolymph of toxic species. We have suggested that it may activate or enhance the smell of the stink glands (Rothschild *et al.*, 1972; Feltwell and Rothschild, 1974) or assist in detoxifying the tissues of species like *Zygaena* which contain HCN (Jones *et al.*, 1962) at all stages of their life-cycle, or even contribute to the toxic substances in the form of degradation products in defensive secretions. Thus Meinwald *et al.* (1968) have shown that a degradation product of neoxanthin is present in the ant-repellent froth of a grasshopper, and Eisner *et al.* (1971) have also demonstrated sesqueterpinoids in the defensive spray from the osmateria of a Papilionid caterpillar (*Battus poly-damus*). This latter observation is particularly interesting since the orange color of this curious eversible gland-structure is due to the presence of carotenoids. Clarke and Sheppard (1973) have described a recessive form of *Papilio memnon* which presumably lacks carotenoids. The larva is blue instead of green in color, which is usual in the absence of these pigments (Gerould, 1921) and the osmateria white. Here again carotenoids obviously fulfill a dual signal - a warning color and chemical repellent.

A fairly large proportion of the Lepidoptera examined so far mirror the carotenoids found in their food plants, both quantitatively and qualitatively. Thus out of 17 carotenoids identified by Feltwell and Valadon (1972) in the cabbage, 14 were found in *Pieris brassicae* reared upon this plant. Furthermore, we found that, fed on a variety of cabbage ("Greyhound") with an unusually high carotenoid content, the concentration in the butterflies was correspondingly high (Table 4). Nevertheless, storage can be selective. The Small White (*P. rapae*), for example, when both weight and size are taken into consideration, stores noticeably less carotenoids than the Large White (Table 5) when both larvae are reared on portions of the same leaves. Those species of Lepidoptera which we have found contain unusually large quantities of carotenoids are aposematic or toxic species with yellow haemolymph and distasteful qualities, such as the Cinnabar (427 μg/g) and the Magpie moth (670 μg/g). There are numerous exceptions (Feltwell and Rothschild, 1974) and the relationship between aposematic species and carotenoid storage clearly required further investigation.

The toxic Large White butterfly (*P. brassicae*) is a highly successful species with an impressive array of adaptations linked with the food plant. Sinigrin in the leaves of the cabbage acts as a feeding stimulus for the larvae, and an oviposition stimulus for the gravid female (Schoonhoven, 1973; Chun and Schoonhoven, 1973). The caterpillar sequesters and stores sinigrin and mustard oils (Rothschild, 1972b) and also large amounts of carotenoids (Table 5). These provide part of the aposematic coloration of the larvae and pupae and the brilliant yellow pigmentation of the eggs. Furthermore, the gregarious caterpillars possess a nauseating odor incorporating the unmistakeable reek of decomposing cabbage. Specimens are fairly easily knocked off the food plant onto the ground, where they are highly cryptic, toning in astonishingly well with the yellowing cabbage leaves and brown soil found in this situation. This is another example of an aposematic species which is cryptic when viewed from a distance, especially when a single individual is accidentally separated from the gregarious assembly.

TABLE 5. Examples of toxic and non toxic British Lepidoptera in which carotenoid storage may be selective (adapted from Marsh and Rothschild (1974) and Feltwell and Rothschild (1974).

Name of species and stage tested	Intraperitoneal injection into laboratory mouse: toxicity	Total carotenoids μg/g	Total carotenoids per insect	Dry weight of insect in gr	Toxins present	Coloration
Large White *Pieris brassicae* adult	Both sexes lethal ♀ : kills mouse in 30 hours ♂ : kills mouse in 4 days	179-647.47 (several trials)	4.50-18.62	0.029	Mustard oils and glycosides	Aposematic at all stages Haemolymph yellow
Small White *Pieris rapae* pupa	pupa kills mouse in 35 hours	66	0.74	0.011	Mustard oils and glycosides	Aposematic in adult stage only Haemolymph yellow
Green-veined White *Pieris napi* pupa	pupa kills mouse in 23-30 hours	24	0.28	0.01	Mustard oils and glycosides	Aposematic in adult stage only Haemolymph yellow
Small Tortoiseshell *Aglais urticae* pupa	not tested	639.10	21.83	0.034	not tested	Aposematic larva and adult
Poplar Hawkmoth *Laothoe populi* adult	negative	17.73	1.77	0.100	100 μg/g histamine	Cryptic (flash display)
Cinnabar Moth *Tyria jacobaeae* larva pupa	not tested kills mouse in 4 days	427.30 304.32	- -	- -	Pyrrolizidine alkaloids Histamine	Aposematic at all stages except pupa Haemolymph yellow
Buff Ermine *Spilosoma luteum* adult ♂ pupa	negative no effect	14.35 -	0.94 -	0.066 -	nil nil	Aposematic (except pupa) Believed to be a mimic of *S. lubricipeda* Haemolymph yellow

Species						
White Ermine *Spilosoma lubricipeda* pupa	30 hours	49.73	1.75	0.035	Histamine (not tested for alkaloids)	Aposematic larva and adult Haemolymph yellow
Gypsy Moth *Lymantria dispar* pupa	negative	61.48	3.25	0.053	not tested	Adult cryptic Larva aposematic
Pale Tussock *Dasychira pudibunda* pupa	not tested	213.33	3.47	0.029	not tested	Aposematic as larva only
Large Yellow Underwing *Noctua pronuba* adult	not tested	25.00	1.91	0.076	not tested	Cryptic (color flash in flight)
The Heart and Dart *Agrotis exclamationis* adult	not tested	21.46	0.77	0.036	not tested	Cryptic at all stages
Six-spot Burnet *Zygaena filipendulae* adult (sex not known)	kills mouse in 12 hours	139.09	'7.55	0.054	HCN, histamine ACh and a toxic protein (?)	Aposematic at all stages Haemolymph yellow
Narrow bordered Five- Spot Burnet *Zygaena lonicerae* adult (gravid ♀)	kills mouse in 2-3 minutes (most toxic species investigated)	-	-	-	HCN, histamine ACh and a toxic protein (?)	Aposematic at all stages Haemolymph yellow
Magpie Moth *Abraxas grossulariata* adult ♀	not tested	676.00	6.75	0.01	histamine-like activity	Aposematic at all stages Haemolymph yellow
pupa	kills mouse in 6 hours					

Particular attention is drawn to the differences between (i) the Large White (aposematic at all stages) and (ii) Small White and Green-veined White butterflies (aposematic as adult only). The first named species is more toxic and contains more carotenoids than the other two. This is not merely a question of size, nor the result of the different feeding habits of the larvae, since we reared the samples tested on parts of the same leaves. An interesting secondary subsidiary observation was made: the diapausing strain of the Green-veined White contains, during the pupal stage, 25% more carotenoids than the non-diapausing strain (Mummery, Rothschild and Valadon, in prep.). Since the caterpillars were fed on the same leaves, one assumes that either the diapausing strain ate more food, or stored carotenoids more efficiently. The color of the diapausing strain is green, irrespective of the background.

DISCUSSION

Natural selection, as we suggested in our introduction, would deal summarily with the random signaller. It is safe to assume that an adequately equipped recipient, capable of arousal and response, must exist if a well developed signal has been evolved. It takes two to make a signaller. But just as in the case of mimicry, which is also a situation grafted on to an existing pattern, preadaptation must play a part in the development of signals.

One of the striking facts about carotenoids is that they are present on such a lavish scale. The annual natural production estimated by Borenstein and Bunnell (1966) amounts to about 100 million tons. It seems necessary for both plants and animals to stockpile these substances and, probably as an insurance policy, to store more than is required immediately. Thus, in the case of various mammals, a considerable amount of carotene passes through the gut wall, which has escaped conversion into vitamin A and is assimilated in fat and other body tissues (Table 3). The golden color of cream and butter in the spring is evidence of the surplus carotenoids ingested and stored by cows. This material is therefore readily available for color signals and relatively little effort is required to push them into the beaks, legs and feathers of birds, and the eggs of warningly colored Lepidoptera and the petals of flowers (Tables 2 and 3). Furthermore, they can achieve the desired effect by very versatile means. The green feathers of various birds are due to the blue structural colors seen through a transparent yellow layer of carotenoids, but the blue, violet and red colors found in the plumage of the oriental fruit pigeons are due to the presence of a red carotenoid which has the property of displaying different colors when absorbed on different substances (Fox and Vevers, 1960, p. 67). Yet again, the contrast of yellow, orange and red in some plumage (i.e., the Goldfinch, Desselberger, 1930) is achieved by a variation in concentration of the same carotenoids. The greenish yellow coloration on the hind wings of the Large and Small White butterflies is due to a subtle combination of blue, black, white and yellow pterines together, with yellow carotenoids in the haemolymph seen through the veins near the dorsal surface of the wings.

Various invertebrates "mobilize" carotenoids into the gonads during the breeding season. A sand crab for example (Gilchrist and Lee, 1972) takes up β-carotene preferentially at this period, and certain crustaceans put 50% of their carotenoids into their eggs. These eggs are sometimes dark green, sometimes colorless, sometimes bright blue. To produce conventional warning coloration must be a relatively simple affair and it is, of course, quite usual to find among arthropods, including the insecta

(Table 3) and molluscs, coral colored, brilliant yellow, orange
and red eggs.

Another feature of carotenoids which preadapts them to use
in a signalling context is their ability to produce attractant
and aposematic coloring on the one hand and cryptic coloration on
the other. Thus the violent reds of humming bird flowers are
ideally suited to attract the pollinators from a distance, while
the yellow haemolymph of ladybirds, oozing round their black leg
joints, is very effective from a few centimeters away. Complexing
with blue bile pigments on the other hand, carotenoids achieve
crypsis for the majority of larval Lepidoptera and, in conjunction
with minute crystals of guanine, which scatter the short waves of
light which strike them, produce the green concealing coloration
of tree frogs.

On the recipient's side all vision and photoreception is
ultimately dependent on carotenoids (Wald, 1960). Quite apart
from the synthesis of vitamin A and the visual pigments, we find
carotenoids concentrated in the fovea of the eye and the colored
oil drops in front of the receptors in the eyes of birds and
turtles, and the yellow intraocular filters in reptiles, amphibia,
some fish and mammals (Walls, 1963; Daw, 1973; Rochon-Duvigneaud,
1972). Yet until quite recent times some doubt existed in the
minds of entomologists regarding the function of carotenoids,
even in insect vision, for as late as 1972 Wigglesworth states
"whether carotenoids play a part in insect physiology is not known.
In mammals, vitamin A is a derivative of carotene, and a related
derivative forms the visual purple of the retina". Only a year
later, however, Dadd (1973) is able to conclude "At least for
normal visual function, exogenous carotenoids are dietary essentials for probably all insects".

This may be only the tip of the iceberg which has loomed up
suddenly out of the mist, and time will reveal an even more
grandiose role than that already envisaged for these golden pigments. All visual information is dependent on carotenoids, but
quite possibly the recipient of signals involving smell, taste
and hearing, as well as sight, may depend ultimately on the
carotenoids handed out to them so liberally by the plant world.

SUMMARY

Since carotenoids are involved in both photoreception and
vision throughout the animal kingdom it is obvious they constitute
one of the most important factors in signal reception. It is
suggested they may also play a part in taste, smell and even
sound reception. Both plants and animals make use of carotenoid
pigments for advertising either attractive or repellent qualities.
Since these substances are available on such a lavish scale many

species are preadapted to develop signals by this means. They are also employed in "dual purpose" signalling. This enables a poisonous or dangerous animal to remain suitably unobtrusive while a potential predator is out of range but to switch rapidly to the display of warning attributes should the animal approach within striking distance. Some examples of high concentrations of carotenoids in toxic insects and plants are discussed.

NOTES

1. Wald (1969) has presented this idea in a brilliant paper in Comparative Biochemistry, in which he points out that "throughout their entire range these excitations appear to be derived chemically from a single closely knit family of compounds, the carotenoids. This relationship persists from phototropism in moulds to vision in man".

2. It is possible that all red fruits and berries attract frugivorous birds and their warning qualities are directed towards other animals. Thus the fruits of *Atropa belladonna* (purplish-black, another avian "favorite" color) are eaten by birds, and are adapted to seed dispersal by this method, although their toxic alkaloids are highly poisonous to mammals. Broad generalizations are possible, like in the case of mimicry in insects, but within this framework every instance may present a special situation which has to be considered on its own merits. Just as there are birds which specialize in eating well protected insects (Rothschild, 1971, p. 218; Rothschild and Kellett, 1972) no doubt there are birds which can eat poisonous fruits with impunity.

3. Quite a few butterflies (*Charaxes*, etc.) clap audibly when they rise vertically in the air and a few use it as a warning signal in fights or chases with other species in their territories.

4. In *Columba livia* male and female use a clapping display.

5. Beetles in semi-desert or arid environments, especially Carabids and Chrysomelids, eat the flowers but also act as pollinators (Verne Grant, 1950, and personal communication).

6. The suggestion that one smells or tastes as well as sees by virtue of vitamin A is not new, but when I first put it forward I naively thought it was. It is not impossible that hearing is sharpened in the same manner. No one has so far explained the pigmentation associated with the organ of corti. Ruedi (1954) claims that intramuscular injections of vitamin A can improve certain types of deafness. So far the yellow pigmented wax in the

external ear of mammals also remains something of a mystery, although its dimorphism indicates its usefulness.

7. Goodwin (1971a) suggests that "massive changes, both morphological and biochemical, are triggered by a ripening hormone" and it is more than possible that a physically separate pathway of carotenoid synthesis arises in developing fruit.

8. The petals of *Crocus sativus* contain a rare water soluble carotenoid (Harborne, 1967). The dried stigmas from which saffron is made contain crocein, a digentiobiosyl ester of crocetin, which provides possibly the sole example of a carotenoid glycoside. The bitter principal is picrocrocin (Weedon, 1971) but, all in all, saffron has an agreeable spicy taste for us. It has been suggested that the yellow variety of the crocus has a sweeter taste than either the purple or white forms.

9. The oxygen carrying properties of vertebrate blood need no comment, but it is probably also of intraspecific importance as a warning color on the case of injured animals.

ACKNOWLEDGEMENTS

I am most grateful to Miss Boothroyd, Dr. J. Feltwell, Mr. B. Gardiner, Dr. N. Marsh, Miss R. S. Mummery, Professor T. Reichstein, Dr. T. Swain, and Dr. G. Valadon for permission to include data from unpublished papers. Mr. Gardiner kindly drew my attention to the special interest in the univoltine strain of *P. napi*. I am also greatly indebted to Professor A. R. Clapham, Professor C. Clarke, Dr. R. W. Crossky, Professor A. Dixon, Dr. D. Eisikovitch, Professor E. B. Ford, Professor T. W. Goodwin, Professor Verne Grant, Professor Howard Hinton, Professor Philip Sheppard, Dr. D. W. Snow and Dr. G. Vevers for discussions (both verbal and by letter) and the loan of many papers. I am also indebted to Dr. S. Y. Thompson for the quotation from Jeremiah.

LITERATURE CITED

Baker, R. R. 1970. Bird predation as a selective pressure on the immature stages of the cabbage butterflies *Pieris rapae* and *P. brassicae*. J. Zool., London 162:43-59.

Bang, B. G. 1971. Functional anatomy of the olfactory system in 23 orders of birds. Acta anat. Supplement 58-1, 79:1-76.

Barbier, M. 1970. Chemistry and Biochemistry of Pollens. *In* Progress in Phytochemistry, Vol. II, (L. Reinhold and Y.

Liwschitz, eds.), pp. 1-34, John Wiley Interscience, London.

Bardach, J. E. and Todd, J. H. 1970. Chemical Communication in fish. *In* Advances in chemoreception, Vol. I, Communication by chemical signals (J. W. Johnston, D. G. Moulton, and A. Turk, eds.), pp. 205-240, Appleton-Century-Crofts, New York.

Baylor, D. A. and Hodgkin, A. L. 1973. Detection and resolution of visual stimuli by turtle photoreceptors. J. Physiol., London (in press).

Bilz, R. 1963. Die Farbe der Nahrungsmittel in anthropologischer Sicht. Erösterungen über die Nahrungsmittel-Ambivalenz. Wiss. veröff. Dt. Ges. Ernähr. 9 (Carotene und carotinoide) 150-167.

Boothroyd, A. and Swain, T. Investigation of the carotenoid content of food plants of aposematic insects. (in preparation).

Borenstein, B. and Bunnell, R. H. 1966. Carotenoids: properties, occurrence and utilization in foods. Advan. Fd Res. 15:195-276.

Briggs, M. H. and Duncan, R. B. 1961. Odour receptors. Nature, London 191:1310-1311.

Butler, C. G. 1970. Chemical Communication in insects: behavioural and ecological aspects. *In* Advances in chemoreception Vol. I. Communication by chemical signals (J. W. Johnston, D. G. Moulton, and A. Turk, eds.), pp. 35-78, Appleton-Century-Crofts, New York.

Chemist and Druggist. 1956. An aid to the identification of colored and marked tablets etc. Ann. Spec. Issue 165(3984): 602-605.

Chun Ma Wei and Schoonhoven, L. M. 1973. Tarsal contact of chemosensory hairs of the Large White butterfly *P. brassicae* and their possible role in oviposition behaviour. Entomol. exp. appl. 16:343-357.

Clarke, C. A. and Sheppard, P. 1973. The genetics of four new forms of the mimetic butterfly *Papilio memnon* L. Proc. Roy Soc. London B 184:1-14.

Dadd, R. H. 1973. Insect nutrition: current developments and metabolic implications. Annu. Rev. Entomol. 18:381-420.

Daw, N. W. 1973. Neurophysiology of colour vision. Physiol.
    Rev. 53:571-611.

Desselberger, H. 1930. Ueber das Lipochrom der Vogelfeder. J.
    Ornithol. 78:328-376.

Ehrlich, P. R. and Raven, P. H. 1965. Butterflies and plants: a
    study in coevolution. Evolution, Lancaster, Pa. 18:586-608.

Eisner, T., Kluge, A. F., Ikeda, M. I., Meinwald, Y. C. and
    Meinwald, J. 1971. Sesquiterpenes in the osmeterial secre-
    tion of a papilionid butterfly *Battus polydamus*. J. Insect
    Physiol. 17:245-250.

Feltwell, J. S. E. 1973. The metabolism of carotenoids in
    Pieris brassicae L. (The Large White Butterfly) in relation
    to its foodplant *Brassica oleracea* var *capitata* L. (The
    Cabbage). Ph.D. thesis, University of London (Royal Hollo-
    way College).

Feltwell, J. and Rothschild, M. 197 . The carotenoids in thirty-
    eight species of Lepidoptera. J. of Zool. Lond. (in press).

Feltwell, J. S. E. and Valadon, L. R. G. 1972. Carotenoids of
    Pieris brassicae and of its food plant. J. Insect Physiol.
    18:2203-2215.

Fox, H. M. and Vevers, G. 1960. The nature of animal colours.
    Sidgwick and Jackson, London.

Fraenkel, G. S. 1959. The raison d'être of secondary plant sub-
    stances. Science 129:1466-1470.

Gerould, J. H. 1921. Blue green caterpillars: the origin and
    ecology of a mutation in haemolymph colour in *Colias
    (Eurymus) philodoce*. J. exp. Zool. 34:385-415.

Gilchrist, B. M. and Lee, W. L. 1972. Carotenoid pigments and
    their possible role in reproduction in the sand crab
    Emerita analoga (Stimpson 1857). Comp. Biochem. Physiol.
    42B:263-294.

Goodwin, T. W. 1950. Carotenoids and Reproduction. Biol. Rev.
    25:391-413.

Goodwin, T. W. 1952. Comparative biochemistry of carotenoids.
    Chapman and Hall, London.

Goodwin, T. W.   1965.   Distribution of carotenoids.   *In* Chemistry
    and Biochemistry of Plant Pigments (T. W. Goodwin, ed.), pp.
    127-173, Academic Press, London and N.Y.

Goodwin, T. W.   1966.   The Carotenoids.   *In* Comparative Phyto-
    chemistry (T. Swain, ed.), pp. 121-137, Academic Press,
    London and N.Y.

Goodwin, T. W.   1971a.   Biosynthesis.   *In* Carotenoids (O. Isler,
    ed.), pp. 577-636, Birkhauser Verlag, Basel.

Goodwin, T. W.   1971b.   Pigments - Arthropoda.   *In* Chemical
    Zoology, Vol. VI. (M. Florkin and B. T. Sheer, eds.), pp.
    279-306, Academic Press, London.

Grant, V.   1950.   The protection of the ovules in flowering plants.
    Evolution, Lancaster, Pa. 4:179-201.

Grant, A. K. and Grant, V.   1968.   Humming birds and their flowers.
    Columbia University Press, New York and London.

Gregory, R. L.   1972.   Eye and Brain.   World University Library,
    Weidenfeld and Nicholson, London.

Harborne, J. B.   1967.   Comparative biochemistry of the flavinoids.
    Academic Press, London and N.Y.

Hinton, H. E.   1973a.   Natural deception.   *In* Illusion in nature
    and art (R. L. Gregory and E. H. Gombrich, eds.), pp. 96-
    159, Duckworth, London.

Hinton, H. E.   1973b.   Some recent work on the colours of insects
    and their likely significance.   Proc. Brit. Ent. nat. Hist.
    Soc. 6(2):43-54.

Hodgkin, A. L.   1971.   Anniversary address: recent work on
    visual mechanisms.   Proc. Roy. Soc. London A. 326:x-xx.

Hopkins, C. Y., Jevans, A. W. and Boch, R.   1969.   Occurrence of
    octadeca-trans-2, cis-9, cis-12, trienoic acid in pollen
    attractive to the honeybee.   Can. J. Biochem. 47:433-436.

Jäger, H., Schindler, O. and Reichstein, T.   1959.   Die Glykoside
    der Samen von *Nerium oleander* L.   Helv. chim. Acta. 42:977-
    1013.

Jones, D. A., Parsons, J. and Rothschild, M.   1962.   Release of hydrocyanic acid from crushed tissues of all stages in the life-cycle of species of the Zygaeninae (Lepidoptera). Nature, London 193:52-53.

Kettlewell, B. 1973. The evolution of melanism. Clarendon Press, Oxford.

Lane, C. D. 1963. Round the blue lamp. Entomologist's mon. Mag. 99:189-195.

Lepage, M. and Boch, R. 1968. Pollen lipids attractive to honey-bees. Lipids. 3:530-534.

Loher, W. 1960. The chemical acceleration of maturation process and its hormonal control in the male desert locust. Proc. Roy. Soc. London B. 153:380-397.

McCartney, W. 1968. Olfaction and odours: an osphiesiological essay. Springer Verlag, Berlin.

Marsh, N. and Rothschild, M. 197 . Aposematic and cryptic Lepidoptera tested on the mouse. J. Zool., London 173 (in (press).

Meinwald, J. Erickson, K., Hartshorn, M., Meinwald, Y. C. and Eisner, T. 1968. Defensive mechanisms of arthropids: XXIII An allenic sesquiterpenoid from the grasshopper *Romalea microptera*. Tetrahedron Lett. 25:2959-2962.

Moncrieff, R. W. 1951. The chemical senses. Leonard Hill, London.

Morère, J-L. 1971. Nutrition des insects. Le carotène: substance indispensable pour la nutrition de *Plodia interpunctella* (Lep. Pyralidae). C. r. Acad. Sci., Paris, Ser. D. 272:2229-2231.

Parmenter, L. 1958. Flies (Diptera) and their relations with plants. London Natur. 37:115-125.

Pasteels, J. M., Deroe, C., Tursch, B., Braekman, J. C., Daloze, D. and Hootele, C. 1973. Distribution et activités des

alcaloides défensifs des Coccinellidae. J. Insect Physiol.
19:1771-1784.

Porsch, O. 1931. Grellrot als Vogelblumenfarbe. Biologia
Generalis 7:647-674.

Portschinsky, J. 1897. (Caterpillars and moths of St. Peters-
burg Province. Biological observations and investigations.)
Lepidopterorum Rossiae biologia. V. Coloration marquante et
taches oscelléses, leur origine et développement (in Russian).
Horae Soc. ent. Ross. 30(1895-1896):358-428.

Proctor, M. and Yeo, P. 1973. The pollination of flowers.
(New Naturalists' Series No. 54), Collins, London.

Reichstein, T. 1967. Cardenolide (herzwirksame Glykoside) als
Abwehrstoffe bei Insekten. Naturwiss. Rdsch. Braunschw.
20:499-511.

Reichstein, T., Von Euw, J., Parsons, J. A. and Rothschild, M.
1968. Heart poisons in the monarch butterfly. Science
161:861-866.

Rochon-Duvigneaud, A. 1972. L'oeil et la vision. In Traité de
Zoologie XVI Mammifères. Fasc. IV. pp. 607-703, Masson et
Cie, Paris.

Rosenberg, B., Misra, T. N. and Switzer, R. 1968. Mechanism of
olfactory transduction. Nature, London 217:423-427.

Rothschild, M. 1961. Defensive odours and mullerian mimicry
among insects. Trans. Roy Entomol. Soc. London 113(5):101-
121.

Rothschild, M. 1964. An extension of Dr. Lincoln Brower's
theory on bird predation and food specificity, together with
some observations on bird memory in relation to aposematic
colour patterns. Entomologist 97:73-78.

Rothschild, M. 1967. Mimicry: the deceptive way of life. Nat.
Hist. N.Y. 76:44-51.

Rothschild, M. 1971. Speculations about mimicry with Henry Ford.
In Ecological genetics and evolution (R. Creed, ed.), pp.
202-223, Blackwells Scientific Publications, Oxford.

Rothschild, M. 1972a. Some observations on the relationship between plants, toxic insects and birds. *In* Phytochemical Ecology (J. B. Harborne, ed.), pp. 2-12, Academic Press, London and New York.

Rothschild, M. 1972b. Secondary plant substances and warning coloration in insects. *In* Insect/Plant Relationships. Symposia of the Roy. Entomol. Soc. London No. 6 (H. F. van Emden, ed.), pp. 59-83, Blackwells Scientific Publications, Oxford.

Rothschild, M., Von Euw, J. and Reichstein, T. 1972. Some problems connected with warningly coloured insects and toxic defence mechanisms. Mitt. Basler Afrika Bibliogr. 4-6: 135-158.

Rothschild, M. and Kellett, D. N. 1972. Notes on the reactions of various predators to insects storing heart poisons (cardiac glycosides) in their tissues. J. Entomol. A 46(2): 103-110.

Rothschild, M., Schlein, Y., Parker, K. and Sternberg, S. 1973. Jump of the Oriental rat flea *Xenopsylla cheopis* (Roths.). Nature, London 239: 45-48.

Rothschild, M., Von Euw, J., Reichstein, T., Smith, D. and Pierre, J. Cardenolide storage in *Danaus chrysippus* (in preparation).

Rüedi, L. 1954. Actions of vitamin A on the human and animal ear. Acta oto-rhino-lar. belg. 5-6: 502-516.

Schoonhoven, L. M. 1973. Plant recognition by lepidopterous larvae. *In* Insect/Plant Relationships. Symposia of the Roy. entomol. Soc. London No. 6 (H. F. van Emden, ed.) pp. 87-99, Blackwells Scientific Publications, Oxford.

Snow, D. W. 1971. Evolutionary aspects of fruit eating by birds. Ibis 113:194-202.

Swynnerton, C. F. M. 1915. Birds in relation to their prey: experiments on wood-hoopoes, small hornbills and a babbler. J. S. Afr. Orn. Un. 11:32-108.

Thompson, W. R. 1943-51. A catalogue of the parasites and predators of insect pests. Belleville Ont. Imp. Paras. Serv.

Tinbergen, N. 1953. The herring gulls' world. Collins, London.

Valadon, L. R. G. and Mummery, R. S. 1968. Carotenoids in flo-
    ral parts of a narcissus, a daffodil and a tulip. Biochem.
    J. 106:479-484.

Valadon, L. R. G. and Mummery, R. S. 1973. A comparative study
    of carotenoids of Ladybirds (Ladybugs) milking aphids
    feeding on vetch. Comp. Biochem. Physiol. 46: 427-434.

Van Der Pijl, L. 1966. Ecological aspects of fruit evolution.
    Proc. Kon. ned. Akad. Wetensch. (Ser. C) 69:597-640.

Von Skramlik, E. 1963. The fundamental substrates of taste. *In*
    Olfaction and taste (Y. Zotterman, ed.), pp. 125-132, Perga-
    mon Press, Oxford.

Wald, G. 1960. The distribution and evolution of visual systems.
    In Comparative biochemistry, Vol. I (M. Florkin and M. S.
    Mason, eds.), pp. 311-345, Academic Press, New York.

Walls, G. L. 1963. The vertebrate eye and its adaptive radia-
    tion. Hefner Publ. Co., Pasadena, California.

Weedon, B. C. L. 1971. Occurence. *In* Carotenoids (O. Isler,
    ed.), pp. 29-59,   Birkhäuser Verlag., Basel.

Weis Fogh, T. 1974. Energetics and aerodynamics of flapping
    flight. *In* Insect flight. Symposia of the Roy. Entomol.
    Soc. London No. 7 (in press).

Wigglesworth, V. B. 1972. The principles of insect physiology,
    7th edition. Chapman and Hall, London.

Witschi, E. 1961. Sex and secondary sexual characters. *In* The
    biology and comparative physiology of birds, Vol. II (A. J.
    Marshall, ed.), pp. 115-168, Academic Press, New York and
    London.

PLATE I

Figure 1. A group of mixed sweets and pills showing how bright
colors are used - rather misguidedly - to arouse two contrasting reactions in man.

Figure 2. Lens of the Zeiss Contarex camera. The small red spot
marks the position in which the close-up lens clicks into
position.

Figure 3. Haemolymph, colored yellow by the presence of carotenoids, oozing out of the defensive glands on the dorsal
surface of the Jersey Tiger *Euplagia quadripunctaria*.

Figure 4. Red mark in the cervical region marking the aperture
of a bleeding point and gland in a Papilionid butterfly.

THE COEVOLUTION OF PLANTS AND SEED PREDATORS

Christopher C. Smith

Division of Biology
Kansas State University
Manhattan, Kansas 66506

INTRODUCTION

The patterns of coevolution of seeds and their predators
has been thoroughly reviewed by Janzen (1971a). Rather than up-
date or criticize that review, I will attempt to classify the
selective forces affecting seeds and their predators in a way
that will help to demonstrate the general causes of the patterns
which Janzen has reviewed. Where selective forces have opposing
effects, I will attempt to demonstrate what information is
needed to define the balance of opposing forces.

GROUPING OF SELECTIVE FORCES

Coevolution is the change in two or more species which are
acting as selective forces on each other. Each species is only
one of a myriad of selective forces acting on the others and its
effect may be countered by several other selective forces.
Instead of a process of change, the interaction between the two
or more coevolving species may often be static. An explanation
for an essentially static situation rests on an understanding of
how the myriad selective forces counterbalance each other. The
explanation can be simplified if the selective forces can be
grouped on the basis of similar patterns of action.

The distinction between selective forces that act as dependent variables and independent variables is one such grouping
that is helpful in explaining why a group of coevolved species
has stabilized in their existing state rather than some other
state that could be imagined. For a selective force to act as
an independent variable it should not be affected by feedback
from the species population upon which it is acting. Many
aspects of climate and geological substrate, or more generally
the physical environment, act in this way. On the other hand,
if one species population such as a predator exerts selection
for more effective defenses in its prey, then the change in the
genetic character of the prey population will feed back as
selection on the predator for more effective means of catching
prey. The same feedback applies for two species competing for
the most effective means of exploiting a common food resource.
As one becomes more efficient at using the food, the decrease in

food available to the other species will select for traits that
make the second species more effective in gaining the common
food. Therefore, the biotic environment acting through competition or predation is a dependent variable. The state at which a
coevolved community stabilizes is ultimately determined by the
nature of the physical environment, while the process of reaching
that state involves dependent feedbacks between the species
evolving together in the community.

The distinction between selective forces acting as dependent
and independent variables should not be confused with the contro-
versy as to whether environmental variables can be divided in a
meaningful way into density-dependent and density-independent
controls of population density (Andrewartha and Birch, 1954).
The point I wish to make is that change in the physical environment is basically independent of biological causes while changes
in the genetic structure of the biotic environment are ultimately
dependent on physical causes.

THE INFLUENCE OF THE PHYSICAL ENVIRONMENT

In many types of coevolved interactions it is extremely
difficult to trace the action of the physical environment, and
the above grouping of selective forces may be of no help. However, for interactions involving seeds, the effect of the
physical environment should be quite evident. Being at the base
of the trophic pyramid, plants are directly dependent on the
physical environment for resources. Being the dispersal stage
and the smallest stage in its life history, seeds have the lowest supply of energy with which to adjust to the physical environment. Selection for packaging the optimum amount of energy
in a seed to allow successful growth of early seedling stages
should be under strong selection from the physical environment.
Strong correlations between seed size and degree of shading of
germination sites (Salisbury, 1942) and between seed size and
moisture stress (Baker, 1972) lend support to the close tie between seed size and the physical environment.

Packaging the optimum amount of energy per seed involves a
compromise between the advantage a parent plant gains from producing a large number of seeds and producing large individual
seeds with a high individual fitness. The optimum compromise can
be represented by a graph plotting individual seed fitness
against individual seed cost (Smith and Fretwell, in press)
(Fig. 1). In the graph, straight lines through the origin serve
as fitness functions for the parents where every point on a line
represents the same number of surviving offspring. The steeper
the slope of the fitness function the greater the fitness for the
parent. A fitness set of the fitnesses for all possible seed

sizes in any given environment should always be convex in shape
(Fig. 1). The fitness set cannot pass through the origin
because the seed must have at least a complete set of genetic
information before it can have greater than zero fitness. The
curve must eventually plateau because there are sites where a
seed might fall, such as a large lake or solid rock, where no
amount of increase in energy would increase the seed's chances of
survival. The point where a straight line through the origin
(parental fitness function) is tangent to the curve of the seed
fitness set is the optimum seed size for the parent to build
because it gives the maximum fitness for the parent. Different
physical environments will give different fitness sets; for
instance a relatively moist environment might give fitness set A
in Figure 1 while a drier environment would give fitness set B.

COMPETITION FOR GERMINATION SITES

In any one area the plant species usually exhibit a broad
range in seed sizes. The trends that Salisbury (1942) and Baker
(1972) describe for the adjustment of seed size to light and
moisture conditions are expressed as differences in the mean
seed sizes of species in different communities and habitats. The
trends are between habitats. They do not explain wide variation
between species within a habitat.

Harper (1965) and Harper, Williams, and Sager (1965) have
demonstrated a basis upon which different species of seeds could
be adapted to different germination sites which exist in a mosaic
within a habitat. The optimum seed size for different germination sites should differ in most cases. In general, smaller
seeds should be able to germinate and grow successfully in a subset of the germination sites that will allow success for larger
seeds (Fig. 2). The graphical arguments for an optimum seed size
used in Figure 1 do not apply to Figure 2 because the ordinate
does not plot seed fitness. In Figure 2 I am representing the
accumulated frequency of a mosaic of germination sites of different physical conditions that will allow seeds of increasing
sizes to germinate. The relationship is likely to be discontinuous as shown in Figure 2. For example, at 1,000 m on the
western slopes of the Cascade Mountains in Washington State the
climax forests contain Pacific silver fir (*Abies amabilis*) and
western hemlock (*Tsuga heterophylla*) whose seeds differ in energy
content by a factor of about 30 (Smith, 1970). The small seeds
of the hemlock are successful in germinating and growing on dead
logs which retain their moisture longer than the litter of
needles and twigs over the rest of the forest floor (Franklin,
1964). Pacific silver fir seeds have enough energy to build a

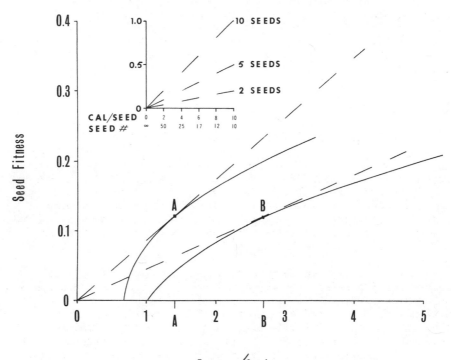

Figure 1. A graphical representation of the interaction
between seed size and seed fitness. The dashed lines
through the origin in the insert and in the main graph
represent fitness functions for the parents of the
seeds. Every point in a fitness function represents
the same number of surviving offspring for the parent.
Thus fitness functions of greater slope represent
higher fitnesses for the parents. The solid curves in
the main graph represent fitness sets for seeds of all
sizes in two different physical environments: A and
B. The fitness sets must leave the abscissa to the
right of the origin because a seed must have the energy
content of at least one set of genetic material before
it has greater than zero fitness. The curves will be

root system for a seedling that will allow it to grow in the
needle litter. Since there is no intermediate between dead logs
and needle litter, there will be a discontinuous shift in the
relationship between energy necessary for success and the per
cent of the area in the habitat allowing success (Figure 2).
Each species should be able to outcompete the other in part of
the area simply because of the energy content of its seeds. In
addition to differences in energy packaging, the use of the energy
may be specialized to differences in germination sites such as
types of fungi available for mycorrhizal relationships and the
mineral solubility in the different types of litter. Competition
between species for germination sites should lead to character
displacement in seed size and specific biochemical adaptation
between species in the same general area. Character displacement
should occur as a result of competition for germination sites
even if the relationship represented in Figure 2 is continuous.

As soon as biochemical or structural differences in seeds
are taken into consideration as being specific adaptations to
germination sites, the use of energy as a general and flexible
resource loses some of its meaning. In order to make generalizations about the effects of selective forces, the function of different chemicals and structures must be defined. The best
definitions are operational ones derived by measuring the relative success of seeds within a species that differ in the
development of the trait in question. Such operational definitions of function are difficult to form for the characteristics
of seeds that affect germination and early growth of the seedling
because of the great difference in the relevant parameters in the
environment of a seed and the environment of the human observer.
It is difficult to "think" like a seed. It is much easier to
form operation definitions for the function of structures

convex and eventually plateau because in any area there
will be some places, such as large lakes or bed rock,
where no amount of increase in seed size will allow
survival. The point at which a fitness function is
tangent to a fitness set is the optimum seed size for
that fitness set because it has the steepest slope of
all the fitness functions intersecting the fitness set.
The less suitable an environment is for the early
growth of a seedling, the further to the right will be
the fitness set for that environment. Thus, the less
suitable environment B will lead to the evolution of a
larger optimum seed size B than the more suitable
environment A.

and chemicals that defend a seed from predation because we discriminate among nuts and seeds on the ease with which they are opened, their taste, and their size during normal feeding. Consequently, when analyzing the interaction of selective forces from the physical environment and those from seed predators, I will simplify the action of the physical environment to its general effect of selecting for the optimum energy content of the endosperm and embryo of the seed.

COEVOLUTION OF SEEDS AND PREDATORS

Having established the general pattern with which seed size evolves in response to the physical environment and interspecific competition, we can look to see whether the selective forces exerted by seed predators act counter to or in concert with the independent variable in the system. There are at least five general patterns with which seed predators can interact with seeds. 1) The predator can discriminate between seed sizes and

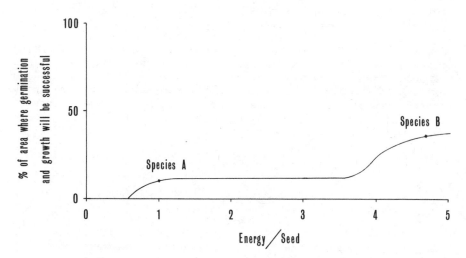

Figure 2. A graphical representation of the relationship between seed size and the percent of the surface of the ground that will allow successful germination and growth of the seedling. Points A and B represent the sizes and the percent of the habitat where germination is possible for two species of seeds that could coexist without competitively excluding each other.

take larger seeds first. This choice could often increase the
foraging rate of an animal that needed more than one seed, or
increase the food supply of an animal that only fed on one seed
in a lifetime. 2) The predator can discriminate between seed
sizes and take smaller seeds first. This choice might be an
advantage to an insect that needed less than the contents of one
seed and could find thinner seed coats to chew through in smaller
seeds. 3) The predator can have no discrimination other than to
choose the correct species of plant to attack, and can successfully attack all seeds. In this case the plants with smaller and
more numerous seeds will have a higher per cent of their seeds
escape predation and therefore be under selection for a decrease
in seed size. Each of these first three patterns will select for
a change in seed size. 4) The predator can have no discrimination other than to choose the correct species of plant to attack,
but some characteristic of the plant other than seed size will
determine whether or not the predator will succeed in killing and
eating the seed. The thickness or texture of seed coats or fruit
may influence the success of the predator's attack. 5) The predator chooses which seeds to attack on the basis of some characteristic other than seed size. In these last two patterns seed size
need not be affected.

In order to determine the extent to which predation affects
evolution of optimum seed size set by the physical environment
there is a need for studies to determine the extent to which
predators operate in the first three patterns as opposed to the
last two patterns. Two studies of complex interactions between
several species of seeds and predators brought rather different
conclusions as to how predators discriminate among food items.
In analyzing the interaction between rodents and conifer seeds, I
came to the conclusion that squirrels (*Tamiasciurus*) operate in
pattern 5 by choosing trees with high numbers of seeds per cone
in which to feed (Smith, 1970). In analyzing the interaction
between bruchid weevils (*Bruchidae*) and seeds of trees in the
Leguminosae, Janzen (1969) came to the conclusion that the weevils
indirectly were selecting for smaller seed size by operating in
pattern 3 with no between-tree discrimination and completely successful attack in some tree species. In neither study was there
an attempt to eliminate the possibility that other patterns of
predator attack were in operation.

Phillip Elliott (in press)   studied the interaction
between red squirrels (*Tamiasciurus hudsonicus*) and lodgepole
pines (*Pinus contorta*) to determine which patterns of attack a
predator uses in influencing the evolution of seeds. His method
was to characterize each of the approximately 300 trees in each
of the territories of three squirrels for all the traits that
might influence squirrel feeding efficiency. Because some

geographic races of lodgepole pines retain seeds in cones on the
tree for many years, he could estimate how many cone crops
remained on the tree uneaten and compare this measure of escape
from predation with characteristics of cones taken from the
trees. A multiple regression of the various cone characteristics
on the number of cone crops remaining on the trees showed that
the squirrels do not discriminate between trees on the basis of
seed size. They do discriminate on the basis of several traits
that make cones difficult to remove from trees and on the number
of seeds per cone. Each of these traits used for discrimination
influences the proportion of the cone that is composed of seeds.
The overall effect of discriminative seed predation by squirrels
is selection for a lower proportion of the trees' reproductive
effort to go into seeds and a higher proportion to go into tissues that protect the seed from predators. The selection has
been pushed so far that only 1% of the weight of cones is in
seeds (Elliot, in preparation). Assuming that organisms
expend the maximum amount of energy on each annual reproduction
which is commensurate with maximizing total future reproduction,
then the effect of seed predation is a reduction in seed number
to allow for an increase in protective devices without a change
in seed size.

The races of lodgepole pines which have closed, or seroti-
nous, cones are adapted to reseeding an area after a fire. They
are found to the east of the Cascade and Coastal Mountain Ranges
in western North America where the forests are seasonally dry and
frequently burned (Critchfield, 1957). In the damp forests along
the Pacific Coast the cones open and shed their seeds during the
autumn of the year they mature. The cones of this coastal race
differ from the interior race in all of the characteristics which
make cones difficult for squirrels to use as food (Critchfield,
1957 and Smith, 1970). In all characteristics they are easier
for squirrels to use than are the cones of the interior race.
The reduction of defensive tissues in cones from 99% to 97%
(Elliot, in preparation) should be a result of the shorter
interval of time the seeds are available in cones for squirrel
attack and therefore a reduction in selective pressure from
squirrels. If squirrels had been influencing the evolution of
seed size by showing a preference for larger seeds, then the
coastal race of lodgepoles would be expected to have larger seeds
than the interior race. However, the coastal race has smaller
seeds (Elliott, in preparation), presumably in response to
the lower level of moisture stress in coastal forests.

The interaction between squirrels and lodgepole pine cones
may be unusual in several ways that make its application as a
general example questionable. There are probably very few
species of seeds that are not regularly attacked by some species

of insect or bird while they are on the parent tree, but seroti-
nous lodgepole pines is one. Therefore, all the variation in
seed mortality on the tree can be attributed to squirrels. Few
species of seeds remain in a predictable position on the parent
plant as long as lodgepole pine seeds. Therefore, lodgepole
seeds have less chance of escaping predators in time and space
than most species. By choosing cones on the basis of the number
of seeds they contain, the squirrels are actually discriminating
on the basis of the edible energy content of the unit for which
they are foraging. If the unit being searched for were a single
seed rather than a fruiting body with many seeds, this type of
discriminative behavior would lead to selection for decreased
seed size.

Although studies of the patterns of discrimination of seed
predators are few, the somewhat artificial interactions between
stored grains and insect pests demonstrate that insects discriminate between seeds of a single species on the basis of hardness
of seeds and seed coats (Davey, 1965; Doggett, 1958; Russell,
1966), texture of the surface of seed coats (Larson, 1927), and
combinations of these traits with seed size (Doggett, 1957;
Russell, 1962; Teotia and Singh, 1966). Future studies of the
coevolution of single species of seeds and seed predators should
be directed at finding the relative importance of predator discrimination of seed size as opposed to difficulty of eating different seeds. Where predators discriminate on the basis of both
types of characteristics, it is important to determine the relative extent to which each type of characteristic can respond to
selection. Does the counterselection from the independent
variable, the physical environment, prevent change in seed size?

Another important consideration in defining the extent to
which predators affect the evolution of seed size, is the consequences of a predator feeding and discriminating among and between the seeds of two species at the same time. This situation
is graphed in Figure 3 for discrimination on the basis of seed
size. A normal distribution of seed sizes for two species is
plotted. The mean size of species B is 4 units, and of species
A, 2 units. The feeding rate of the seed predator increases
linearly with seed size if one assumes that the main time consuming activity is finding the seed or extracting it from fruit
or seed coats for large seed predators, or laying an egg on it
for small seed predators. If chewing seeds were the main time-
consuming activity, then feeding rates would not be proportional
to seed size, and the graphical arguments would not be applicable.
Figure 3 demonstrates that if predation rates are linearly proportional to seed size, then the per cent of the day needed for
feeding or finding oviposition sites increases at a geometric
rate with decreasing seed size. Fitness gained from discrimina-

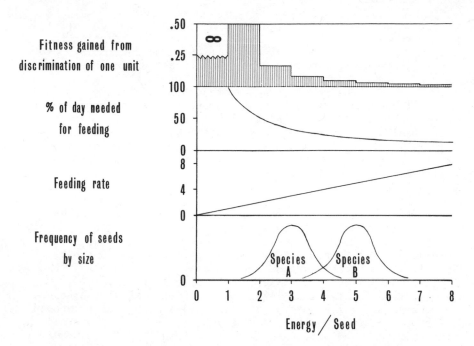

Figure 3. A graphical representation of the relationship between seed size and the feeding efficiency of a seed predator. The bottom graph shows the frequency distribution of seed sizes for two species of seeds, A and B. The second lowest graph shows the feeding rates possible for a seed predator using different sizes of seeds. The third graph shows the relationship between seed size and the per cent of the day a predator would need for feeding on seeds of different sizes. The top graph shows the gain in fitness of a predator that results from its distinguishing between seeds that differ by one unit of energy and taking the larger. For each bar in the graph, the seed sizes being distinguished are those on either edge of the bar. As argued in the text, a predator would be most likely to maximize its feeding efficiency if it ate all the seeds in species B first in an indiscriminant manner and then discriminated between seeds in species A on the basis of their size.

tion should be proportional to the time gained from the discrimination. Figure 3 shows that discrimination of one unit of energy
between two different seed sizes gives increasingly larger gains
in fitness with decreasing seed size. The consequences of the
relationships shown in Figure 3 are that a predator will gain
more by discriminating among the seeds of species with small
seeds than among the seeds of a species with larger seeds. Further, the predator will gain more by discriminating between the
two species than it will by discriminating among the seeds of the
larger species. Since the sensory activity necessary for discrimination between species is likely to require less effort from
a predator than to discriminate among the seeds within a species,
it is likely that a predator will first discriminate between the
seeds of two species and attack the seeds of the larger species
in an indiscriminate manner before it finally attacks the seeds
of the smaller species. The effect of this pattern of attack
will be the absence of directional selection on seed size in the
larger species. This was indeed the pattern demonstrated by
squirrels while feeding on the seeds of conifers (Smith, 1970).
The smaller species could still experience selective pressure for
smaller seed size, but the larger species can only respond to
selective pressure from seed predators by change in some characteristic that makes individual seeds more difficult to find or
extract. In conifers, the larger species of seeds are shed from
the cones earlier in the fall and are available for squirrel
predation a shorter period of time than smaller species of seeds.
In general, it is likely that predators exert selection for
smaller seed size only in the smallest of the species of seeds it
is feeding on in a discriminative manner.

The interaction of selective pressures from the physical
environment and from seed predators would be greatly simplified
if the former acted mainly on the energy content of the endosperm
and embryo of the seeds and the latter acted mainly on the extent
and nature of the tissues surrounding the seeds. As yet, there
is very little evidence to support or refute this pattern of
interaction of selective pressures. Future studies of the coevolution of seeds and their predators should be designed to gain
such evidence.

SEED PREDATORS AS DISPERSAL AGENTS

In some instances seed predators are also the chief dispersal
agent of the seeds upon which they feed (Janzen, 1971a). This
relationship is found where animals store, or scatter hoard, seeds
in the ground for future use. Some of these seeds germinate
before they are rediscovered. The relationship complicates the
interaction between seed and seed predator because in order for

the seed to be dispersed it must be desirable enough for the
predator to wish to store it and eat it. An animal that is
strictly a seed predator acts as a selective force against characteristics that make seeds more efficient to use. But a predator that scatter hoards seeds exerts counterbalancing selective
forces both for and against these traits.

The importance of considering the desirability of a seed to
a predator that scatter hoards is demonstrated by the interaction
of two types of seeds with either gray squirrels (*Sciurus caroli-
nensis*) or fox squirrels (*S. niger*) in the deciduous hardwood
forests of eastern North America (Smith and Follmer, 1972). The
acorns of white oaks (*Quercus alba*), bur oaks (*Q. macrocarpa*),
and shumard oaks (*Q. shumardii*) have thin shells and a low lipid
and protein content when compared to the nuts of shagbark hickory
(*Carya ovata*) and black walnut (*Juglans nigra*). These differences
between oaks as opposed to hickories and walnut are apparently
family differences and will apply to most, or all, comparisons
between the Fagaceae and the Juglandaceae. In some areas of
eastern North America the climax communities are mixed stands of
oaks and hickories. In order for the seeds to be effectively
dispersed in these communities both types must be desirable to
squirrels which are the main scatter hoarders in the area. Table
1 shows that there are consistent differences in the food value
of the two types of seeds. Because acorns have thinner shells,
they can be eaten faster and supply calories of metabolizable
energy at a faster rate than hickory nuts or walnuts. However,
because hickory nuts and walnuts have a higher caloric value per
gram dry weight and are digested with a greater efficiency than
acorns, a squirrel can obtain the same amount of food energy with
less weight in its stomach by eating hickory nuts. In the fall
when squirrels are active in storing food and in the spring when
they are mating, the physical exertion of these activities is
probably easier when carrying a lighter load in the digestive
tract and the longer feeding periods needed for hickory nuts and
walnuts are no disadvantage in the mild weather of those seasons.
In winter, when squirrels spend a large amount of time in the
insulation of their nests, the faster feeding rate allowed by
acorns would be an advantage in conserving body heat and the
heavier weight in the digestive tract would be no disadvantage in
the nest. Nixon *et al.*, (1968) found the peak period of acorn
use to be in the late fall and winter and the peak period of
hickory nut and walnut use to be in the spring and fall for gray
and fox squirrels in Ohio. In mixed stands of oaks and hickories,
a squirrel would have to scatter hoard both types of nuts to
insure an optimum winter and spring food supply, and would serve
as an effective dispersal agent for both types of seeds.

The selective forces affecting oaks and hickories is greatly

Table 1. Energy content and digestibility of five species of nuts from trees of the deciduous hardwood forests of eastern North America.

Tree species	Kcal /gram dry weight of nut kernels (Kcal/ gram dry weight)	% of nut's energy in the kernel (%)	Metabolizable energy per gram dry weight eaten (Kcal)	Metabolizable energy eaten per minute (Kcal/minute)
White oak	4.17	27.4	3.02	1.18
Bur oak	4.34	45.0	3.60	2.48
Shumard oak	5.22	65.9	4.49	2.18
Black walnut	6.23	12.7	5.63	0.75
Shagbark hickory	6.57	21.2	6.16	1.07

complicated by the numerous other types of mammals, birds, and
insects that feed upon them. The information that Follmer and I
gathered demonstrates that squirrels discriminate between nuts on
the basis of their nutritional value and the speed with which
they can be consumed. However, the differences in the biochemical composition of the kernels of the two types of nuts may well
be related to the specific requirements of their germination
sites. The effect of the various seed predators on the two types
of seeds may be to increase the proportion of the reproductive
effort that goes into protective tissue in the more nutritious
hickory nuts and walnuts as compared to acorns (Table 1). Squirrels could keep the selective pressures affecting the two types
of nuts in balance, but until more is known about the other types
of seed predators, the effect of squirrels must remain conjecture.

Smith and Bradford (in preparation) describe another interaction between seeds and the predators that scatter hoard them.
In this case something is known about the action of all the important selective forces and a rough description of their interactions will give an indication of the pattern of their balance.
The seeds of the palm (*Scheelea rostrata*) in Central America
develop in a complex fruit about 3 to 5 cm long (Figure 4). The
fruits are composed of a fibrous and impermeable exocarp which
protects the surface of the fruit while it is growing and when it
first falls to the ground. Beneath the exocarp is a fleshy and
oily mesocarp which is eaten by several species of birds and mammals (Janzen, 1971b; Smythe, 1970; Fleming, 1971). The function
of the mesocarp, which has the same caloric value as the seed,
seems to be to supply nutrition to the animals at the time the
seeds are being scatter hoarded. Below the mesocarp is a very
horny and thick endocarp which surrounds the endosperm and embryo
of the seeds. The endocarp is thick enough and hard enough that
Sciurus granatensis gains less food energy per unit time than
gray squirrels of comparable size do from any of the nuts in the
deciduous forests of eastern North America. The slow feeding
rate allowed by the endocarp could give added advantage to producing a mesocarp so that rodents could feed quickly and still
have time to move nuts away from below the parent tree. Wilson

Figure 4. The structure of *Scheelea* palm fruits and the
patterns of holes made in their endocarp by various
predators. Part 1 is a longitudinal section through a
whole fruit with the point of attachment to the tree at
the left. Part 2 is a cross section of the whole fruit;
e & e = the endosperm and embryo of one seed, en =
endocarp, g. p. = germination pore, mes = mesocarp, and
ex = exocarp. Part 3 is a cross section of the endocarp

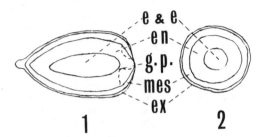

e & e
e n
g. p.
mes
e x

1   2   3   4

 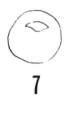 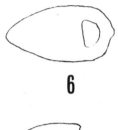

5   6   7

 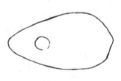

8   9   10

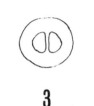

11

or nut containing two seeds. Part 4 is the cross section of a nut with three seeds. Parts 5 and 6 are lateral views rotated 90° from each other of a nut opened by rodents on BCI. Part 7 is an end view of the nut in 5 and 6. Parts 8 and 9 are lateral views of nuts opened by rodents at Taboga. Part 10 is an end view of 9. Part 11 shows the round exit hole made by an emerging bruchid.

and Janzen (1972) have demonstrated that scattering the nuts
reduces their attack from bruchid weevils, which is to the advantage of both the tree and the rodent which wishes to use the nuts
during time of food shortage at the end of the wet season. A
peculiarity of these palm fruits is that the endocarps from the
same tree may contain either 1, 2, or 3 seeds (Figure 4; 2-4).
Further, the trees near the Rio Higueron on Finca Taboga, Guana-
caste Province, Costa Rica (Taboga site) have a significantly
higher frequency of nuts with 2 seeds (23%) and 3 seeds (4.0%)
than do the trees on Barro Colorado Island, Panama Canal Zone
(BCI site) which have 5% of their nuts with 2 seeds and 0.2%
with 3 seeds.

The fitness of the fruits with 1, 2, and 3 seeds can be
broken into 4 components for each type of fruit. The first component is the energy cost of building each of the three types of
fruits (Table 2). The energy cost is estimated by caloric content
of the tissues in the fruit. Although this estimate does not
include the different amounts of respiratory energy used in
building different tissues, it is a comparison of proportions of
the same four tissues in each fruit. The errors should not be as
great as would result from comparing different types of tissues
in different parts of a plant. The fitness of energy cost for
the three types of fruits is inversely proportional to the energy
content of each type of fruit, since the numbers of fruits a tree
can build is inversely proportional to the size of an individual
fruit. We assume that fruits with 2 or 3 seeds can only give
rise to one adult tree because the endocarp is indehiscent and
two adult trees, each with a trunk 25 cm in diameter, would not
have room to grow from a single nut.

The second component of fitness is predator escape (Table 3).
Survival of at least one seed from predation is highest for nuts
with 3 seeds (79%), next highest for nuts with 2 seeds (64%), and
lowest for nuts with 1 seed (39%). This pattern results from
predators killing some but not all of the seeds in nuts with 2 or
3 seeds. Relative fitness of the three types was calculated by
dividing each of the frequencies of survival by the highest (79%).

The third component of fitness is the increase in survival
resulting from dispersal. This component was estimated from the
relative amount of mesocarp available in the three types of fruit
with which to attract dispersal agents. Since some seeds are
successful in germinating and growing for a while under the
parent tree, it was estimated that dispersal doubled a seed's
probability of reaching adulthood rather than being an absolute
prerequisite for survival. To calculate the fitness value in
Table 3 we divided the caloric content of the mesocarp of the
type of fruit with the most mesocarp into the caloric content of
the mesocarp in each type of fruit. The ratios give a comparison

Table 2. Energy content (Kcal) and caloric value (Kcal/gram dry weight) of four different tissues in *Scheelea* palm fruits of three different types.

Number of seeds per nut	Exocarp (Kcal)	Mesocarp (Kcal)	Endocarp (Kcal)	Endosperm and embryo (Kcal)	Endosperm and embryo in 1 seed (Kcal)	Total energy in 1 fruit (Kcal)
1	6.9	12.3	36.3	7.2	7.2	62.7
2	9.0	13.8	51.0	11.8	5.9	85.7
3	7.6	10.0	49.0	16.9	5.6	84.4

Caloric value (Kcal/gram dry weight)

	4.40	7.64	4.67	7.72		

Table 3. Relative fitness values (W) for 4
components of the relative fitness
of three types of *Scheelea* palm
fruits.

Number of seeds per nut	W for energy cost	W for predator escape	W for dispersal	W for competitive ability
1	1.00	0.50	0.95	1.00
2	0.73	0.80	1.00	0.81
3	0.74	1.00	0.86	0.73

of the relative attractiveness of the three types of fruits for
dispersal agents. Then we took half the difference between each
fitness and 1.0 and added that to the fitness to account for
doubling the probability of survival rather than having dispersal
an absolute prerequisite for survival.

The fourth component of fitness is the competitive ability
that accrues from the size of a single seed in each of the three
types of fruit. This component was calculated by assuming that
each of the three types of fruits have the same overall relative
fitness and that all the major components of fitness other than
competitive ability of the seedling in its germination site were
already taken into consideration. The fitness for competitive
ability was that number which when multiplied with the other
three components of fitness would give the same overall fitness
for each type of fruit. When fitness for competitive ability is
plotted against seed size it gives a convex fitness set (Figure
5) as was predicted earlier. If fruits with one seed had seeds
as small as a single seed from fruits with 2 or 3 seeds, then the
parent of those smaller fruits would have a lower fitness than
the parent of the existing fruits with one seed. The curve in
Figure 5 does not preclude there being larger seed sizes which
intersect a line through the origin with a steeper slope than the
tangent in Figure 5. That possibility could not be tested.

The calculation of competitive fitness of seedlings is based
on the assumption that all three types of fruit have the same

fitness. The assumption is supported by the discriminatory behavior of the rodents which scatter hoard the endocarps containing
seeds. Both rodents and bruchids attack all three types of nuts
with equal frequency. However, rodents are very discriminatory
in determining that a nut they have attacked has a second or
third seed in it. In the Costa Rican population, nuts with two

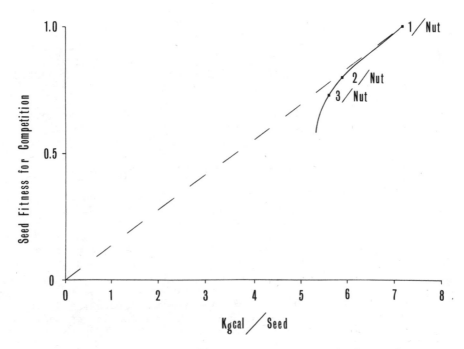

Figure 5. The relationship between seed size and a
seed's fitness for producing a seedling of varying
competitive ability as determined for single seeds
from *Scheelea* palm fruits that contain 1, 2, or 3
seeds. Seeds from fruits with one seed would give
their parent the highest fitness as shown by the
parental fitness function intersecting the point
1/nut.

seeds that were attacked by rodents had only 12% with one surviving seed while in nuts with two seeds that were attacked by
bruchids 51% had one surviving seed. The evidence for bruchids
discovering a second seed in a nut was inconclusive. Since they
attack a nut by laying an egg on the surface of the endocarp,
there are very few sensory clues for them to use in discovering a
second seed. Rodents, however, should be able to sense the difference in the thickness of the walls of a nut from which they
have removed one of two or three seeds.

    That the discriminatory behavior of rodents maintains a
balance in the overall fitness of the three types of fruits is
supported by three lines of evidence. 1) Nuts with 2 and 3 seeds
are more frequent in Taboga than on BCI as mentioned earlier. 2)
Nuts are attacked more frequently by bruchids (52%) than rodents
(10%) at Taboga and more frequently by rodents (63%) than bruchids (18%) on BCI . 3) Rodents attack nuts in a way that
exposes two or three seeds most of the time at Taboga (Figure 4;
8-10) while rodents attack nuts in a way that usually exposes
one seed on BCI (Figure 4; 5-7). The method of opening nuts in
the region of the germination pore (Figure 4; 1) used by rodents
at BCI should require much less chewing effort than the method
used at Taboga. It appears that rodents are able to discern
second and third seeds in a nut, but only do so when these nuts
occur at a relatively high frequency in the population. The
probability that an individual rodent will systematically search
for a second seed in a nut probably increases with the frequency
of the nuts with two seeds much as in the search images formed by
other vertebrates (Croze, 1971). On BCI where rodents are the
chief predators the advantage gained in predator escape that
comes from nuts with two and three seeds is lost when a few
rodents start to form search images for second seeds. At Taboga
where rodents perform a smaller fraction of the predation, the
advantage of nuts with two and three seeds is not lost until most
of the rodents are forming search images for second seeds. The
maintenance of the frequencies of nuts with 1, 2, and 3 seeds is
performed by frequency dependent predation of second seeds by
rodents coupled with the relative importance of rodents and bruchids as predators. The frequency dependent predation of second
and third seeds maintains a uniform overall fitness for the three
types of nuts.

    Again in this example of coevolution of seeds and seed preda-
tors, it is necessary to understand the power and extent of the
discrimination used by the seed predators in order to understand
the balance reached in the system. There is some evidence that
the independent variable in the system is the physical environment working through seed size. In Figure 5 the fitness set of
seed size shows that an 18% decrease in seed size leads to a

selective coefficient of only .01 against the parent producing
the smaller seed. The relationship is shown by drawing straight
lines from the origin through the points for 1 seed per nut and 2
seeds per nut. The slope of the line through 2/nut is only 1%
lower than the line through 1/nut and the slopes of the lines are
a measure of parental fitness. There would be relatively little
counter selection from the physical environment to resist selective pressures exerted by predators for smaller seed size for at
least a decrease of 18%. If predators had selected for smaller
seed size, they would have pushed seed size to a steeper region
of the curve of the fitness for seedling competitive ability.

It is likely that the general effect of seed predators is to
select for an increased thickness of endocarp rather than to affect
seed size. Janzen (1971b) indicates that the larvae that hatch
from some bruchid eggs on the surface of a nut are unsuccessful
in chewing through the endocarp to reach the seed. The endocarp
could be under selection from bruchids to be thick enough and
hard enough to prevent some of the bruchid larvae from penetra-
ting. Rodents have the power of sensory discrimination to distinguish the thin area of the endocarp near the germination pore.
It is likely that they could discriminate between the nuts of
different trees on the basis of the ease of chewing away the
endocarp to expose the seeds. Thus the method of attack by both
bruchids and rodents would allow patterns of selection that would
effect distribution of energy into protective tissue without
influencing the energy content of the endosperm and embryo.

The second means of predator defense is the polymorphism in
numbers of seeds per nut. In this defense the advantage gained
in avoiding predation by having 2 or 3 seeds in a nut is countered
by a decrease in the size of individual seeds. However, seed
size is not the major change in morphology between the three
types of nuts. The single seeds in nuts with 2 and 3 seeds are
82% and 78% of the energy content of the seeds in 1-seeded nuts,
rather than 50% and 33% as might be expected if seed size were
not under strong selection for the optimum size. Packaging the
surprisingly large seeds in nuts with 2 and 3 seeds is accom-
plished by a 17% and 23% increase in the diameter of the nuts and
a reduction of the thickness of endocarp over the seeds to 82%
and 79% of that found in nuts with one seed.

The polymorphism in seed number also influences the success
of scatter hoarding by rodents as a means of seed dispersal. The
fact that some nuts with 2 and 3 seeds are opened by rodents
without all the seeds being killed allows some nuts that are dispersed and later rediscovered and attacked to still give rise to
seedlings. The polymorphism in types of nuts has a specific
advantage for being scatter hoarded.

CONCLUSIONS

The coevolution of seeds and seed predators can be best understood if the selective forces that affect species populations are separated into those that act as independent variables and those that act as dependent variables. Both the amount of energy in endosperm and embryo that is used in the early growth of a seedling and the amount of energy in tissues that inhibit predator feeding efficiency could be affected by selective pressure from either the physical environment (independent variables) or from seed predators (dependent variables). However, it is likely that the energy content of endosperm and embryo is affected mainly by the physical environment and the energy content of tissues that inhibit feeding of seed predators is affected mainly by selective forces exerted by the seed predators. Although three studies have been described as supporting the above conclusion, only the study of squirrels and lodgepole pines by Elliott (in press)   clearly eliminates other possible interpretations.

The separation of the function of tissues in the endosperm and embryo from the function of tissues in surrounding seed coats and fruit is an oversimplification. Specific chemicals in the endosperm and embryo may inhibit predation (Janzen, 1969 and 1971a) while seed coats may influence the specificity of germination sites and the success of the early growth of seedlings (Harper, 1965; Harper, Williams and Sager, 1965). The only valid definitions of the functions of chemicals, tissues, structures, or behavioral patterns are operational definitions. Studies of the coevolution of seeds and their predators must include measurements of the variation in the characteristics of the individual populations under investigation and must demonstrate how different components of the environment affect the relative reproductive success of individual organisms which differ in the characters whose function is to be defined.

SUMMARY

This paper attempts to classify and analyze the types of selective pressures that are involved in the coevolution of seeds and their predators. The selective forces are divided into those that act as independent variables and as dependent variables in the evolution of a stable community structure. Generally the physical environment acts as an independent variable on the evolution of seeds by selecting for an optimum energy content of endosperm and embryo to support the early growth of seedlings. Seed predators act as a dependent variable which mainly influences the development of protective tissues in and around the seed and does

not effect the size of the endosperm and embryo. Three examples
of the discriminative behavior used by seed predators in attacking seeds are given to support this general pattern of action of
selective forces as dependent and independent variables.

ACKNOWLEDGEMENTS

I wish to thank David F. Bradford, Phillip F. Elliott, and
Stephen D. Fretwell for the use of as yet unpublished information.
Doris Marx and my wife Ann were of great help in preparation of
this manuscript.

LITERATURE CITED

Andrewartha, H. G. and Birch, L. C. 1954. The distribution and
    abundance of animals. University of Chicago Press, Chicago.

Baker, H. G. 1972. Seed weight in relation to environmental
    conditions in California. Ecology 53:997-1010.

Critchfield, W. B. 1957. Geographic variation in *Pinus contorta*.
    Maria Moors Cabot Foundation Pub. no. 3. Harvard University
    Press, Cambridge.

Croze, H. 1971. Search image in carrion crows. Zeitschriften
    für Tierpsychologie beiheft 5:1-85.

Davey, P. M. 1965. The susceptibility of sorghum to attack by
    the weevil *Sitophilus oryzae* (L.). Bull. Entomological
    Research 56:287-297.

Doggett, H. 1957. The breeding of sorghum in East Africa. I.
    Weevil resistance in sorghum grains. Empire J. Exper. Agric.
    25:1-9.

Doggett, H. 1958. The breeding of sorghum in East Africa. II.
    The breeding of weevil-resistant varieties. Empire J.
    Exper. Agric. 26:37-46.

Elliott, P. F. (volutionary responses of plants to seed-eaters:
    pine squirrel predation of lodgepole pine. Evolution (in
    press).

Elliott, P. F. Phenotypic variation in lodgepole pine (*Pinus
    contorta*) (in preparation).

Fleming, T. H. 1971. Population ecology of three species of neotropical rodents. Misc. Pub. Mus. Zool. Univ. Michigan 143:1-77.

Franklin, J. F. 1964. Ecology and silviculture of the fir-hemlock forests of the Pacific Northwest. Proc. Soc. Amer. Foresters.

Harper, J. L. 1965. Establishment, aggression, and cohabitation in weedy species. *In* The genetics of colonizing species (H. G. Baker and G. L. Stebbins, eds.), pp. 243-265, Academic Press, New York.

Harper, J. L., Williams, J. T. and Sager, G. R. 1965. The behavior of seeds in soil. I. The heterogeneity of soil surfaces and its role in determining the establishment of plants from seeds. J. Ecology 53:273-286.

Janzen, D. H. 1969. Seed-eaters versus seed size, number, toxicity, and dispersal. Evolution 23:1-27.

Janzen, D. H. 1971a. Seed predation by animals. Ann. Rev. Ecol. and Systematics 2: 465-492.

Janzen, D. H. 1971b. The fate of *Scheelea rostrata* fruits beneath the parent tree: predispersal attack by bruchids. Principes 15:89-101.

Larson, A. O. 1927. The host-selection principle as applied to *Bruchus quadrimaculatus* Fab. Ann. Entomol. Soc. Amer. 20: 37-79.

Nixon, C. M., Worley, D. M. and McClain, M. W. 1968. Food habits of squirrels in southeast Ohio. J. Wildl. Mgmt. 32:294-305.

Russell, M. P. 1962. Effects of sorghum varieties on the lesser rice weevil, *Sitophilus oryzae* (L.). I. Oviposition, immature mortality, and size of adults. Ann. Entomol. Soc. Amer. 55:678-685.

Russell, M. P. 1966. Effects of four sorghum varieties on the longevity of the lesser rice weevil, *Sitophilus oryzae* (L.). J. Stored Prod. Res. 2:75-79.

Salisbury, E. J. 1942. The reproductive capacity of plants; studies in quantitative biology. G. Bell and Sons, London.

Smith, C. C. 1970. The coevolution of pine squirrels (*Tamias-ciurus*) and conifers. Ecol. Monogr. 40:349-371.

Smith, C. C. and Bradford, D. F. Effects of differential seed predation on *Scheelea* palm nuts (in preparation).

Smith, C. C. and Follmer, D. 1972. Food preferences of squirrels. Ecology 53:82-91.

Smith, C. C. and Fretwell, S. D. The optimal balance between size and number of offspring. Amer. Natur. (in press).

Smythe, N. 1970. Relationships between fruiting seasons and seed dispersal methods in a Neotropical forest. Amer. Natur. 104:25-35.

Teotia, T. P. S. and Singh, V. S. 1966. The effect of host species on the oviposition, fecundity and development of *Callosobruchus chinensis* Linn. (Bruchidae: Coleoptera). Bull. Grain Technology 4:3-10.

Wilson, D. E. and Janzen, D. H. 1972. Predation on *Scheelea* palm seeds by bruchid beetles: seed density and distance from the parent palm. Ecology 53:954-959.

ANT-PLANT MUTUALISM: EVOLUTION AND ENERGY

Brian Hocking

Department of Entomology
University of Alberta
Edmonton, Alberta, Canada   T6G 2E3

Insect plant relationships have been classified (Southwood, 1973) into six groups; those in which the plant provides the insect with food, shelter, or transport, and those in which the insect provides the plant with these assists. Any one relationship may of course fall into more than one group. It seems desirable to add two further groups, in which either partner provides the other with assistance towards reproduction (Table 1). Southwood (1973) has also assessed the relative importance of these six groups for insects in general.

ANT-PLANT MUTUALISM

Ants differ from most other insects in their relationships with plants principally in the restriction of their power of flight to the sexual forms, in which it facilitates reproduction (of the colony) and transport for ants themselves. Thus the major importance of insects in general for the transport and reproduction of plants is denied to the ants. Its place is taken by a major role in the sheltering of plants, though protection or even defense is often a more apt term. This is a reciprocal function; that is, sheltering of ants is also a major contribution of plants to this mutualism, matching in importance, perhaps, their food contribution (Table 2).

Plants as ant food:

Like other social insects, ants have apparently progressed from an animal towards a plant diet. Their ancestors are believed to have preyed on a wide range of insects, other arthropods and some other invertebrates. They probably supplemented this diet with such plant products as sap, nectar, and excretions of plant sucking insects. Some ants still have catholic tastes. Others have specialized for prey on particular groups of insects including other ants, or other arthropods, sometimes on particular developmental stages. Most have incorporated an increasing proportion of materials of plant origin into their diets, again with varying specificity. Unspecialized plant tissue rarely forms more than a small portion of the diet of an ant. Sap, nectar and nectaries, more especially those outside of the flowers, other specialized

Table 1.   The relative importance of insect-plant
           interaction (modified from Southwood,
           1972).

	Shelter	Food	Transport	Reproduction
Plants for Insects	**	***	*	*
Insects for Plants	*	*	**	***

tissues commonly held to be produced by plants as "ant-bait," and
seeds are the more important items. Unspecialized plant tissue,
and sap are however very widely fed upon indirectly, through the
agency of fungi-leaf tissue, of Homoptera-sap, or termites-woody
tissue.
     The fungi used by leaf-cutting ants to convert their harvest
to acceptable form are not specific to the plant, but are very
specific to the ant, being rarely transferrable outside the genus.
They are grown on a prepared medium consisting predominantly of
leaf tissue, selected largely for its availability, faeces of the
culturing species of ants, often supplemented by collected caterpillar faeces - which are mainly residues of leaf tissue. The
identification of these fungi is peculiarly difficult since the
formation of reproductive structures has apparently been selected
against by the ants' culture methods. Yet the ants can clearly
recognize them since they use their mandibles to mechanically
weed out the hyphae of alien species. Leaf cutting fungus growing
ants are confined to the New World.
     The relationships of ants with Homoptera are usually rather
less specific than those with fungi, but still quite specific.
Some species of Homoptera have only been recorded with a single
species of ant; most ant species have been recorded with more
than one species of Homoptera. Homoptera themselves may be quite
specific in their plant hosts. The most important food which ants
obtain from Homoptera is the sweet fluid faecal material known as
"honey-dew" which in some associations may provide the entire food
of the ant. More usually the bodies of Homoptera surplus to those

Table 2. The relative importance of
ant-plant interactions.

	Shelter	Food	Transport	Reproduction
Plants for Ants	***	***		
Ants for Plants	***	*	*	*

needed for the supply of honey-dew are also eaten (Way, 1963).

Harvesting ants, which store plant seeds are characteristic of desert or semi-desert areas, or at least areas with a pronounced dry season. The choice of seeds is not specific, and the effect on plant distribution is minor.

Several species of termites provide another ant-plant go between, for generalized and woody plant tissue. A number of ant species are semi-specific predators on termites, carrying out periodical raids on their nests.

In addition to general plant tissue and floral and foliar nectar, many plants with specific mutualistic associations with ants produce special nutritive structures, rich in fats and protein (F. Darwin, 1877). These may be sterile stamens, Müllerian bodies - spheroidal structures among dense hairs at the base of the petiole, or Beltian bodies - modified tips of leaflets. Such structures are usually distinctively colored.

Plants as ant shelters:

Primitively ants are soil insects and nest in the soil; but the roots of plants are also in the soil, their rigidity aids in excavation and they may support Homoptera. It is reasonable to suppose that ant nests moved above ground by this route, so that nesting in plants and especially trees, because of their permanence, was a natural outcome. It is thus not surprising that so many tree nesting ants have a relationship with Homoptera.

Some ant species will nest in almost any tree cavity resulting from rot, cracks, or accident. Almost any plant, usually

perennial, which has either natural cavities such as hollow stems
or pockets of pith or other soft tissue may be used for nesting
by members of a variety of ant species in an opportunist manner.
Of much greater interest to us are the more regular associations,
often between a specific ant and a specific tree, on an obligate
mutualistic basis. These associations fall into two habitat
groups, both tropical: those of dry, open savannas and those of
wet tropical forests. The former are typified by the relationship between *Pseudomyrmex ferruginea* F. Smith and the leguminous
Acacia cornigera L. in eastern Mexico (Janzen, 1967) and between
Crematogaster mimosae Santschi and *Crematogaster nigriceps* Emery
and *Acacia drepanolobium* (Harms ex) Sjöstedt in East Africa
(Hocking, 1970). Those of wet tropical forests include much more
diverse groups of both plants and of ants. Among them are the
flacourtiaceous *Barteria fistulosa* Masters of West Africa with
its *Pachysima aethiops* (F. Smith), the many species of the rubiaceous *Myrmecodia* with equivalent species of *Iridomyrmex* of the
Malay peninsula, East Indies, and N. Queensland, and many species
of the moraceous *Cecropia* of the New World tropics occupied
principally by species of the dolichoderine genus *Azteca*. Relatively little has been added to our knowledge of the ant plants
of the wet tropics since the studies of Bequaert (1922) and
Wheeler (1942).

The nature of the relationship between these "myrmecophytes"
and their ants has been a controversial one since Hernandez (1651)
first wrote about them. Two conflicting opinions have been current: that the ants exploit the trees and that the association
is mutually advantageous. Nobody, I believe, has ever suggested
that the trees exploit the ants. The ants enjoy an enviable
reputation for nous and success with man and presumably with
other species. The basis for this is dubious; Wilson (1971,
citing Williams, 1964 and Brown, personal communication) suggests
that their approximately 13,000 species, which represent perhaps
one three-hundredth of the total number of insect species, comprise only perhaps one-thousandth of the total insect population.
Both Janzen's (1967) examination   with the Mexican and my
(Hocking, 1970) observations on the East African *Acacia* associations indicate clear advantages to the ants comprising almost
ideal conditions of shelter, especially for the brood, and a
balanced and almost complete diet for all stages. Janzen considered that the impermeable walls of the swollen thorns of *A.*
cornigera favored both the maintenance of an adequate humidity in
the dry season and the exclusion of rain and hence control of
fungus in the rainy season. He also found a temperature excess
of 1 to 3°C in isolated thorns above that of the environment. In
East Africa in *A. drepanolobium* temperature excesses up to 18°C
in insolated small black swellings were measured; a rough correla-

tion of swelling color with altitude, from white at sea level to
black at 1,200 m, was also noted within the ant bearing *Acacia*
spp. White galls heated up in the sun less than black, and large
less than small.

Despite the intricate specificity of their relationship with
Acacia swellings, these *Crematogaster* take readily to alternative
sites. When two large swellings from different trees, each containing queens and brood, were left on the laboratory bench over-
night, one party raided the other, which moved its brood into an
inverted 1 pint aluminum mug on the draining board of the sink
where workers were caring for it, using the runnels of the
draining board as exits and entrances.

Perhaps the most remarkable way in which ants use trees for
shelter is by pulling the edges of neighboring leaves together
and using half-grown larvae to sew them thus with the silk they
secrete. Related ants use this silk to spin a coccoon at pupa-
tion. A colony may use a hundred or more of these pendulous leaf
bags, distributed over several trees, replacing them as the leaves
wither. Leaves of many different species of tree are used in
this way. The most important group of weaving ants is the genus
Oecophylla, restricted to the Ethiopian, Oriental, and Australasian regions (Sudd, 1963, 1967).

Ants as plant protectors:

It is difficult to understand how anybody who has encountered
an ant-acacia in the field in E. Africa could doubt the efficacy
of the ants in protecting the trees. Sometimes even when a tree
is approached and always as soon as it is touched, the ants pour
out of their holes in the swellings and scramble towards the ends
of the branches, emitting an odor which most people find repul-
sive, while adopting an attitude with the abdomen erect which
even a formicophile can recognize at once as threatening. It has
been argued that this is in defense of the brood; but how can the
brood be protected without protecting the tree which protects and
feeds it? This prompt aggressive response to any kind of inter-
ference is apparently characteristic of mutualistic plant ants.

The swellings used by ants on most African acacias are the
modified bases of stipular thorns. Unlike the New World species
there is a discrete demarcation between the swelling and the rest
of the thorn, the swellings of the two thorns often coalesce.
Initially they are green, soft, and filled with pith; the color
progressively changes, ultimately to usually black or white.
Before the outer layers harden up they are usually penetrated by
ants at the junction of the unmodified part of one or other thorn
with the swelling. The pith is chewed away from the wall and
often disappears, but some species retain it as a loose matrix

within which eggs and young brood are dispersed and supported.
Scales, used by the ants for honey-dew, are attached when present
where the swelling lies against the stem, which is the only part
of it which is vascular. The large (up to 10 cm) black or white
swellings are conspicuous objects and may well serve as sign-
stimuli for the mature thorns. The young green thorns are soft
and present no problem to ungulates so that the growing shoots
are vulnerable. Some ant species maintain aphid colonies on
them, all take great interest in them, possibly awaiting new
foliar nectaries or usable swellings and certainly compensate for
the softness of the young thorns. Thorns, young and old, are no
protection against herbivorous insects.

Ants as pollinators:

Ants are relatively hairless, lack specific pollen collecting
devices, and are unable to move rapidly from plant to plant.
These features all militate against effective plant pollination.

Despite these unpromising qualities the contribution of ants
to the pollination of *Theobroma* is of economic importance. They
may also be important to cashew nuts (*Anacardium occidentale* L.),
capsicum (*Capsicum frutescens* L.), lychee (*Litchi chinensis* Sonn.)
(Free, 1970), and presumably to other plants of no economic
importance.

Most of the pollination done by ants must be self-pollination,
and much of it may be accidental. Probably this is why the floral
nectaries of many plants are protected from the visits of ants by
sticky surfaces, leaf held water, barriers of hairs, or recurved
slippery leaf edges, thus preventing competition with cross pollinating bees (Lubbock, 1929).

The difficulty in persuading the fungi which are cultured by
leaf-cutting ants to form reproductive structures implies that the
ants make the necessary arrangements for their reproduction, by
vegetative means. This situation has presumably been arrived at
by selection by the ants, and of course has resulted in fully
obligate mutualism.

Ants as plant disseminators:

It is mainly the harvester ants, genera such as *Messor* and
Pogonomyrmex that, by their collection of seeds in dry areas,
contribute to the dissemination of plants. They are ground nesting species and store the seeds in special cells. Some species
with a broad distribution including wetter habitats only store
seeds in the drier parts of their range. Harvester ants are
apparently able to control the germination of the seeds they collect and also exert considerable influence on the vegetation in

the vicinity of the nest. Seed for storage is collected largely
in relation to its availability, with some tendency to prefer the
seeds of grasses and perhaps compositae.

Although seeds from a disused nest may germinate, it is
doubtful whether plants are often dispersed ever an important
distance in relation to other means of seed dispersal.

Ants as plant food:

No clear line of demarcation can be drawn between plants
which protect their flowers from ants and other undesirables, and
insectivorous plants which capture, digest, and utilize such insects as nourishment. It is an interesting fact and worthy of
further study that predatory insects (and sometimes other predators) are to be found with unexpected frequency in the traps of
insectivorous plants. Whether these animals are all attracted by
the same mechanism, or whether, as seems likely, the predators
are attracted by potential prey among earlier victims of the
predatory plant is uncertain. Whichever way it is, ants show up
in appreciable numbers in the traps of insectivorous plants.

EVOLUTION AND ENERGY

Ants and Acacias:

More than half of the World's known species of *Acacia* are
Australian; despite this Bentham considered the genus to be more
tropical (Oriental) in origin and regarded the Australian flora
as a secondary proliferation. No ant associations with *Acacia*
are known from either of these regions, despite the richness of
their ant faunas. In both the Ethiopian and Neotropical regions
there are about 75 species of *Acacia* and in each region about 15
of them have associated ant species. It seems that it was only
when the genus extended its range into more challenging territory
that the protection of ants became necessary. In Africa the
emphasis of the challenge was more on browsing ungulates, in the
New World it was more on herbivorous insects (perhaps in particular, leaf-cutting ants).

Although both foliar nectaries and Beltian bodies tend to
maintain a good distribution of ants over the leaves, there are
usually only two nectaries to a leaf as compared with several
hundred Beltian bodies. Beltian bodies should thus be incompar-
ably more effective in compelling the ants to provide the sort of
coverage needed to protect against other insects. The larger
African swellings with more color contrast on the other hand, provide the sort of visual impact required to deter vertebrates.

I have postulated (Hocking, 1970) that the swellings of the

stipular thorns of African *Acacia* originated as galls of Homoptera, probably aphids. Those of some species have a gross structural resemblance to the galls of some species of *Pemphigus* on
North American poplars, to which there are also histological
resemblances. The sensitivity of the plant to this initial
stimulus would have been strongly selected for by virtue of the
great advantages conferred by the ants. If the threshold of
response became low enough it might then have led to the mere
presence of any Homoptera, and finally perhaps any arthropod,
resulting in this response. The diversity of swelling form
between the species of *Acacia* then becomes comprehensible in
relation to the diversity of species of Homoptera associated with
both the *Acacia* in the first place, and also with the ants. This
explanation is comparable with the "Baldwin effect" or genetic
assimilation used to explain the callosities on newly hatched
ostriches (Eaton, 1970).

Although when grown from seed in the greenhouse, with minimal exposure to arthropods of any kind, a few plants of *A. drepanolobium* failed to produce any swellings after nearly two years,
these were the smallest trees. In nature such plants are not
found - presumably they are eliminated by ungulates.

That coevolution of *A. drepanolobium* and its two associated
species of *Crematogaster* has been in progress for a long time is
clear from the number and diversity of ant guests and other
closely associated insects found and the remarkable resemblances
between them and both the *Acacia* and the ants (Fig. 1). Janzen
(1966) has demonstrated that the same is true of the New World
ant-acacia associations.

The abundance of *A. drepanolobium* and its two species of
Crematogaster and the scarcity of any of the three separately is
clear *prima facie* evidence of the success of the combination and
hence of the truly obligate nature of the mutualism.

In the New World the modification of the thorns of both
Acacia cornigera and *A. sphaerocephala* Schlecht and Cham. does
not appear to be sufficiently profound to call for special explanation. The greater the radius of the thorns the more ants could
be accommodated and the greater the selective advantage. The
occurrence of variously twisted thorns, however, seems to reflect
the same sort of plasticity of thorn form as the African species
exhibit.

Energy budget of *Crematogaster mimosae*:

Just an an inadequate energy budget means the death of an
organism, it may also mean the extinction of a species. Energy
relationships are thus an important consideration in evolution.
Nectar and related liquids such as honey-dew, which are basically

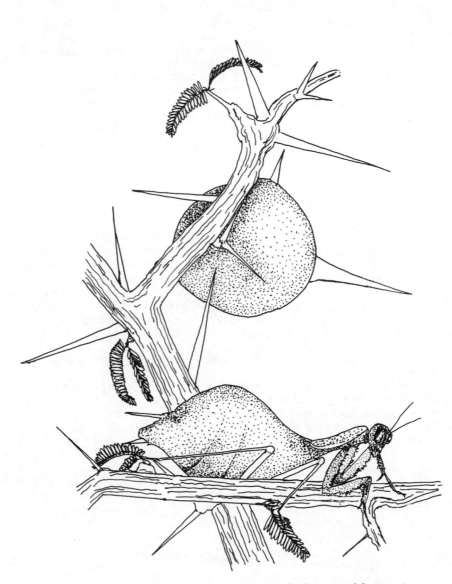

Figure 1. Remarkable resemblance of the mantid *Spnodro-mantis obscura* to the stipular swellings of *Acacia drepanolobium* indicates long and intimate association between this insect and ant plant. *Crematogaster* ants which inhabit this *Acacia* are major prey items for the mantid.

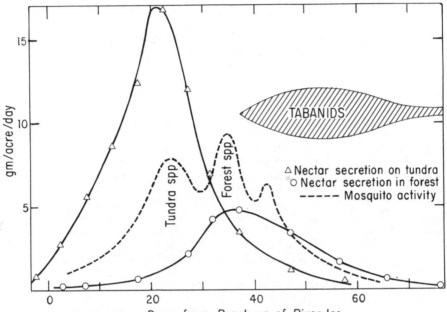

Figure 2. The seasonal changes in nectar production
and biting fly populations, Churchill, Manitoba, 1947-
1952. (Modified from Hocking, 1953.)

sugar solutions, play a tremendous role as energy sources for perhaps a majority of adult endopterygote insects - nearly all
Lepidoptera, most Hymenoptera and Diptera, and many Coleoptera.
The volume and concentration, and hence the energy value, of such
solutions are readily measured in the field with wax-lined micro-
pipettes and a pocket refractometer. A wider use of these simple
techniques would contribute much to our knowledge of energy relationships.

This technique was used to estimate the total sugars (as
glucose) available daily to the average ant population of an
average tree of A. drepanolobium from its foliar nectaries and

scale insects inside the swellings (Hocking, 1970). The result
came to foliar nectaries 3.0 + scales 1.07 = 4.5 g. The total
activity of the ants was computed from two counts each hour of
the numbers moving up and down on the trunk over a 24 hour period
and the energy required for this was added to that for 24 hours
of resting metabolism, to give a figure in terms of glucose per
day of 6.5 g. That is to say, about 65% of the energy needs of
the ants could be met from these sources alone. This budget
would be vastly changed by the production of a mating flight;
flight is an energy expensive activity.

Nectar secretion and insect flight:

It is interesting to compare this nectar production of 3.08 g
per tree per day, which with a tree density of about 500 per hectare gives 1540 g/hectare/day from one species of tree near the
equator with that produced in total by 11 of the most productive
plant species on the tundra at Churchill, Manitoba, Canada, at
latitude 58° N over the whole productive season of about 40 days
at similar insolation, namely 1540 g/hectare (Hocking, 1953).
The corresponding figure for Lake Hazen, Ellesmere Island, latitude 82° N is 1020 g/hectare (Hocking, 1968). At these high latitudes, as Figure 2 shows, flight of some groups of insects at
least is restricted to the period of nectar secretion. Total
nectar yield at high latitudes seems to represent about a half of
one per cent of total productivity while these data for E. African
savanna suggest a much higher percentage for the tropics.
    The use of the same techniques for measurement of sugar solu-
tions permit a minimum estimate of the flight distance of which a
particular insect is capable, for species which feed exclusively
on sugar solutions; for example, mosquitoes from female swarms at
Churchill were found to be carrying on the average enough sugar
solution in the ventral diverticulum to allow a flight of about
25 km (Hocking, 1953). It also allows the nectar of particular
plants or the honey-dew of particular insects to be translated
into equivalent flight distances for insects inbibing them; thus
the nectar produced per hectare of tundra at Churchill in a season is equivalent to about 1.4×10^8 mosquito-km of flight, the
honey produced by an average beehive is equivalent to about 6 x
10^9 mosquito-km, one catkin of *Salix arctophila* Cockerell provides for some 950 mosquito-km daily.

Thermal energy and the evolution of heliotropism:

The absorption of solar radiation by the "black body" surface
of many acacia swellings in the tropics makes an important contribution, even there, to the heat budget of the ants within, at

least at moderate elevations above sea level. Such effects might
be expected to be much more important at high latitudes where the
short season means that every calorie must be made to count. Two
observations at Lake Hazen, Ellesmere Island, 82° N latitude led
us to one example of this (Hocking and Sharplin, 1965). In photo-
graphing flowers of *Dryas integrifolia* Vahl. and of *Papaver radi-
catum* Rottb. it was found to be impossible to take a front view
of the flowers against the light; in observing *Aedes nigripes*
Zett. feeding on the nectar of *Dryas*, both males and females were
noted to spend several times as long resting in a flower as they
needed to feed to repletion. The first observation indicated
that the flowers were heliotropic, constantly turning to face the
sun; the second indicated that the insects were getting something
else besides nectar and their black color and the paraboloid
shape of the flowers suggested that they were enjoying the principal foci of thermal solar radiation. Measurements with thermo-
couples showed that this was so, and gave temperature increments
up to 4.4°C. It must be supposed that this heliotropic mechanism
and paraboloid flower form has been evolved by the plant as a
device ensuring the ripening of its own germ cells in this rigorous environment, and that the insects get theirs matured too, as
well as the meal of nectar. Perhaps the enforced proximity to
the germ cells of the plant contributes to pollination, by these
or other insects.

SUMMARY

The interactions between insects and plants are classified
into eight categories, and the relative importance of these in
the relationships between ants and plants are indicated. Those
most closely involved in the evolution of obligate mutualisms
between ants and plants, that is, those relating to food and
shelter, are given particular attention. Attention is drawn to
the importance of energy needs in these interactions and in
insect-plant interactions generally. The role of sugar solutions
in these energy needs is emphasized.

LITERATURE CITED

Bequaert, J. 1922. Ants in their diverse relations to the plant
    world. Bull. Amer. Mus. Nat. Hist. 45:333-585.

Darwin, F. 1877. On the glandular bodies on *Acacia sphaero-
    cephala* and *Cecropia peltata* serving as food for ants. J.
    Linn. Soc. (Bot.) 15:398-409.

Eaton, T. M. 1970. Evolution. Norton, New York.

Free, J. B.   1970.   Insect pollination of crops.   Academic Press, London and New York.

Hernandez, F.   1651.   Rerum medicarum Novae Hispaniae thesaurus seu plantarum, animalium et mineralium Mexicanorum historia. Rome, 14: +950 + 22 + 90 + 6 pp.

Hocking, B.   1953.   The intrinsic range and speed of flight of insects.   Trans. R. ent. Soc. Lond. 104:223-345.

Hocking, B.   1968.   Insect-flower associations in the high Arctic with special reference to nectar.   Oikos 19:359-388.

Hocking, B.   1970.   Insect associations with the swollen thorn acacias.   Trans. R. ent. Soc. Lond. 122:211-255.

Hocking, B. and Sharplin, C. D.   1965.   Flower basking by Arctic insects.   Nature (Lond.) 206(4980):215.

Janzen, D. H.   1966.   Coevolution of mutualism between ants and acacias in Central America.   Evolution 20:249-275.

Janzen, D. H.   1967.   Interaction of the bull's-horn acacia (*Acacia cornigera* L.) with an ant inhabitant (*Pseudomyrmex ferruginea* F. Smith) in eastern Mexico.   The University of Kansas Science Bulletin 67:315-558.

Lubbock, Sir John.   1929.   Ants, bees and wasps.   London and New York.

Southwood, T. R. E.   1973.   The insect/plant relationship - an evolutionary perspective.   *In* Symp. R. ent. Soc. Lond. No. 6 Blackwell (van Emden, ed.).

Sudd, J. H.   1967.   An introduction to the behaviour of ants. Arnold, London.

Way, M. J.   1963.   Mutualism between ants and honeydew-producing Homoptera.   Ann. Rev. Entomol. 8:307-344.

Wheeler, W. M.   1942.   Studies of neotropical ant-plants and their ants.   Bull. Mus. Comp. Zool. 110: No. 1 Cambridge.

Williams, C. B.   1964.   Patterns in the balance of nature. Academic Press, London and New York.

Wilson, E. O.   1971.   The insect societies.   Belknapp Press, Harvard.

COEVOLUTION OF ORCHIDS AND BEES

Calaway H. Dodson

Marie Selby Botanical Gardens
800 S. Palm Avenue
Sarasota, Florida 33577

Orchids and their pollinators have been extensively discussed in recent years (see Dodson, 1967 for review). In a few instances, such as the euglossine bee-orchid example, the relationships between pollinators and the orchids which they visit have been fairly well elucidated (Dodson et al., 1969; Dressler, 1968a and b; Evoy and Jones, 1971). In many other cases, hints of considerable complexity in those relations have resulted from brief observations of pollinators visiting orchid flowers (van der Pijl and Dodson, 1966). Many of those interesting relationships seem to stem from the adaptation to pollination mechanisms by the orchids based on deception of the pollinator. Pseudocopulation, pseudoprey, pseudoterritoriality, false nectaries and stamens, pseudofragrance and simple color and form convergence on food plants in the biome, abound in the orchids. More than half of the orchid species are believed not to provide food in the customary sense for their pollinators (Dodson, unpublished).

In most of these instances the orchid species have clearly developed floral form, color and fragrance which make it possible for the flowers to interject themselves into a point in the life cycle of their pollinator in order to accomplish their fertilization. It is assumed that the evolutionary force involved is a means whereby highly specific pollinators may be attracted. The extreme is that reported by Nierenberg (1972) where sympatric, compatible populations of *Oncidium bahamanse* Nash and *O. lucayanum* Nash on Grand Bahama Island are reproductively isolated by *O. bahamense* attracting male bees of *Centris versicolor* Fabricius while *O. lucayanum* attracts female bees of the same species, *Centris versicolor*. Flowers of *O. bahamense* imitate enemy insects to be driven away by the male bees who defend their territories while flowers of *O. lucayanum* appear as food producing flowers of *Malpighia glabra* Linn (Malpighiaceae) to the female bees. The two sympatric orchid species have proven to be interfertile when artificially cross-pollinated but natural hybrids between the two have not been reported.

It is hard to imagine coevolution occurring between orchid and pollinator in such cases. The pollinator has more the status of a dupe. However, the disadvantage must be slight or the victim would likely adapt to avoidance of the interaction.

ORCHIDS AND GOLDEN BEES

One of the examples of orchid-pollinator relations which
have been worked out to a relative degree of completeness is that
of the euglossine bee-euglossine orchid. Reports of euglossine
bees visiting orchids date back to Darwin (1862) but the special
nature of the visitation was not reported until the early 1960's
(Dodson and Frymire, 1961; Vogel, 1963; Dressler, 1967). Only
recently has the activity of the male bees, while visiting orchid
flowers, been recognized as lek behavior (Dodson, unpublished).

The occurrence of lek type formations in the male euglossine
bees, commonly known as "golden bees," may have played an important role in the rich development of much of the remarkably
diverse flora and fauna of the American tropics. The word lek is
a term which has been applied to the curious phenomenon in which
certain spectacular male birds such as Birds of Paradise, Hummingbirds and Manakins gather to display their magnificent colors to
each other in characteristic rituals. These displays attract
female birds of the same species, and mating takes place. In the
bird species which have this behavior the male birds do not share
the duties of nest construction, egg incubation or feeding of the
young. The formation of leks appears to be a means of assuring
mating between animals where one of the sexes has a vagabond life
style.

The lek phenomenon is similar among males of the euglossine
bees, a group of brilliantly colored bees found only in the new
world tropics. Immediately after emerging from the communal
nests, the nomadic males leave and never return. Flying freely
through the tropical forests, they feed on nectar producing
plants and sleep hanging from the underside of a leaf or in a con-
venient flower. The female bees collect nectar and pollen along
with nest construction materials and are closely associated with
the nest site.

The male bees are commonly found visiting orchid flowers and
flowers of certain other plants which do not produce nectar, pollen or any other obvious food (Fig. 1 and 2). Landing on the
flowers, they brush the surface with special pads on their front
feet, launch themselves into the air and squeeze the contents of
the pads into cavities in their swollen hind legs. They then
return to the flower and repeat the process, which could go on
for hours. Actually, they are collecting floral fragrances in
their hind legs as borne out by gas chromatographic analyses
(Dodson et al., 1969). Each type of flower visited has a characteristic fragrance spectrum with each kind of bee collecting fragrance components from only a few flower species (Dodson and
Hills, 1966; Hills, Williams and Dodson, 1968; Dodson, 1970).

It was originally believed that the male bees collected the

Figure 1. A male bee of *Eulaema nigrita* brushing on the surface of the lip of *Cycnoches aureum* and collecting fragrance components.

Figure 2. A male bee of *Eulaema cingulata* collecting fragrance components from the spadix of *Spathiphyllum cannaefolium* (Araceae).

fragrance components in very particular combinations in order to attract only females of their own kind. Experiments with the fragrance compounds both in pure form and in the floral combinations proved, however, to attract only male bees (Williams and Dodson, 1972). Although several reports of male golden bees doing curious flight and buzzing rituals were originally attributed to territoriality, it was found that these rituals attract rather than repel other male bees of the same species. The result of this behavior is a swarm of male bees on a small scale. Rarely are more than five male bees involved in one swarm though several swarms may be in progress within a few yards of each other. The female bees are attracted to the swarm, probably by the accumulation of brilliantly colored bees, and mating takes place. Recognition of proper species is probably a close range visual or tactile process. Each species of male euglossine bee has a characteristic pad of short hairs at the base of the tibia of the middle leg. The species can be most easily distinguished on the basis of the form of these pads. Swarms of these male golden bees seem to serve the same purpose as, and are very similar to, the lek behavior in birds.

Recent developments have made it possible to **artificially produce** these leks. Dressler (personal communication) reported in 1966 that bees were occasionally attracted to dead bees of their own species, even when pinned and placed in insect boxes and set out in the tropical sun to dry. Experiments in which legs of male bees of a single species were pinned to blotter pads and set out confirm this report (Dodson and Evoy, unpublished). Individual pads numbered 1 to 30 were set out at the Rio Palenque Biological Station in Ecuador and on each pad was pinned one of each pair of hind legs (the other was saved for gas chromatographic analyses) of male bees of *Eulaema cingulata* (Fab.). The bees had been obtained for the experiment as they were attracted to (but not permitted to collect from) blotter pads saturated with eugenol.

Pads 9 and 17 received frequent visitation from other male bees of *E. cingulata*. Pads 4, 6, 15 and 28 were infrequently visited. The other pads were ignored (see Table 1). On the second day the order of the pads was scrambled to avoid a position effect but the results were about the same. The indication was clear that the legs of bees 9 and 17 were highly attractive while 4, 6, 7, 15 and 28 were only slightly attractive. The other bee legs did not attract. Gas chromatography of the legs indicated that bees 9 and 17 had large quantities of volatile compounds while bees 4, 6, 7, 15 and 28 had low amounts. The other legs had little to no volatiles stored. It is assumed that bees 9 and 17 had had time to collect abundant floral fragrance compounds in appropriate combinations while the other bees had not.

Table 1. Frequencies of attraction of living male bees of
Eulaema cingulata to legs of male bees on the same
species attached to blotter pads numbered 1 to 30
on day 1 and day 2.

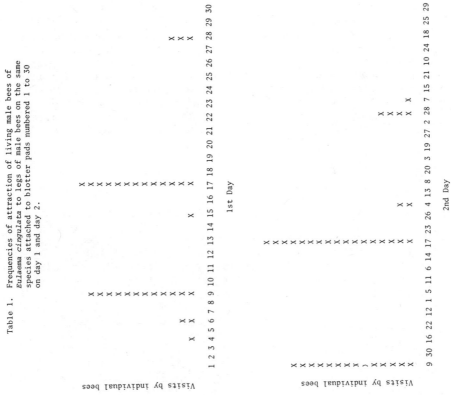

The attraction of male bees to leks was presumed to be on the basis of fragrance attractants; however, other potential attractants such as sound from their loud buzzing or visual cues could not be discounted. Dodson and Evoy (unpublished) made extensive recordings of particular species of bees while buzzing at attractants at Cerro Campana in Panama. The recordings were played back at varying intensities and no attraction was obtained. There is no evidence that bees can sense airborne vibrations (Michener, personal communication). We did succeed in scaring a female three toed sloth with baby down from a Cecropia tree, however.

By chance it was found that tethered male euglossine bees sometimes attract large numbers of male bees of their own species. In February 1973 a male bee of *Eulaema cingulata* was accidently caught in a mist net set up at the Rio Palenque Biological Station in Ecuador for the collection of birds. Fifty-seven male bees of *E. cingulata* were then captured (prior to collecting fragrances) by putting out blotter pads soaked with eugenol. Though the bees had not had eugenol available previously, some of the bees smelled strongly of floral fragrance. The bees were tethered with thread and hung 2 meters apart along a forest trail. Bees number 22 and 37 proved to be strong attractors while bees number 3, 12, 21 and 27 were weak attractors. The other bees did not attract. At the beginning the bees were active and buzzing loudly. In order to determine whether the sound was a factor, the wings of bee number 22 were removed. Attraction continued. The bee still made a slight buzzing sound so the head was removed. Attraction continued. Bee number 37 was placed in a paper bag to test visual attraction. The bag had a few tiny holes punched in it. Though the bee was quiescent and not visible to other bees, attraction continued.

Obviously, most of the captured bees did not have the proper spectrum of naturally collected and stored fragrance compounds to attract other males. In some species, such as *Euglossa mixta* Friese and *Euglossa tridentata* Moure, as many as 300 males had to be captured before a bee with all of the necessary components to be an attractor was encountered. Once an attractor is encountered he can be hung up at will, dead or alive, and other male bees are attracted, sometimes in incredible abundance.

It is presumed that the leks are established only by attractors. When a bee has gathered sufficient quantities of the proper spectrum of floral fragrance compounds, he then establishes a lek. Gas chromatographic analysis of such bees revealed large quantities of up to 14 distinguishable compounds present in the leg. Other male bees of the same species captured at flowers or in leks proved to have few to no components and were not attractive to other bees.

CONCLUSIONS

     In those orchids pollinated by euglossine bees it seems
probable that coevolution has occurred. The plant species which
are adapted to these bees have evolved different floral fragrances
to attract different pollinators. This has made possible extensive speciation in the Orchidaceae, Araceae and Gesneriaceae and
members of certain other plant families in the Neotropics. By
the same token, the presence of numerous, easily collected,
natural plant chemicals probably effected the development of the
bee species. Approximately 60 distinct floral fragrance compounds have been isolated and about 30 of them have been tentatively identified (Williams and Dodson, 1972; Dodson, unpublished). Each plant species which produces them in the flowers
has characteristic combinations of the fragrance compounds which
act to screen the kinds of bees attracted (Williams and Dodson,
1972). There are usually several species of golden bees (up to
40 in one location in Panama) found in any given place in neotropical forests. Frequently only one bee species is attracted
to a particular species of plant. Though the same bee may be
attracted to other kinds of plants, the other plants are either
mechanically or genetically isolated.
     The bees are dependent on the presence of flowers which pro-
duce easily collectable fragrance components. The bees collect
these in order to attract other male bees of their species and
form leks where mating takes place. Relatively minor changes in
chemical composition of floral fragrances could upset the attraction system. The bees may be sensitive to different fragrance
composition from population to population. Evidence exists to
substantiate individual differences in attraction potential of
different floral fragrances for bees of apparently the same
species in the same bee population. For example, thousands of
male *Eulaema cingulata* have been collected or observed while
being attracted to eugenol, vanillin and benzyl acetate without
any visits to cineole, available nearby. Two male bees of *E.
cingulata*, strongly attracted to cineole, were captured in Ecuador
among a normal population of *E. cingulata* which were not attracted
to cineole. *Euglossa viridissima* Friese is much more frequently
attracted to methyl cinnamate in northeastern Mexico while the
same species is more strongly attracted to eugenol in western
Mexico when both compounds are offered. The bees could develop
ethological isolating mechanisms on the basis of distinctive
floral fragrance components presented by the flowers which they
visit.

SUMMARY

Most orchid species have had little effect on the evolution of their pollinators. The euglossine bee pollinated orchids may be an exception since the orchids provide floral fragrance components which are collected by the male bees. The male bees utilize the floral fragrance components to form leks where mating takes place. Changes in sensitivity on the part of the bees could result in effective reproductive isolating mechanisms.

LITERATURE CITED

Darwin, C. 1862. The fertilization of orchids by insects. London.

Dodson, C. H. 1967. Relationships between pollinators and orchid flowers. Atlas do simposio sobre a Biota Amazonica 5:1-72.

Dodson, C. H. 1970. The role of chemical attractants in orchid pollination. Biochemical coevolution. Oregon State University Press 83-107.

Dodson, C. H. and Frymire, G. P. 1961. Natural pollination of orchids. Mo. Bot. Gard. Bull. 49:133-152.

Dodson, C. H., Dressler, R. L., Hills, H. G., Adams, R. M. and Williams, N. H. 1969. Biologically active compounds in orchid fragrances. Science 164:1243-1249.

Dodson, C. H. and Hills, H. G. 1966. Gas chromatography of orchid fragrances. Amer. Orch. Soc. Bull. 35:720-725.

Dressler, R. L. 1967. Why do euglossine bees visit orchid flowers? Atlas do simposio sobre a Biota Amazonica 5:171-180.

Dressler, R. L. 1968a. Observations on orchids and euglossine bees in Panama and Costa Rica. Rev. Biol. Trop. 15:143-183.

Dressler, R. L. 1968b. Pollination by euglossine bees. Evolution 22:202-210.

Evoy, W. H. and Jones, B. P. 1971. Motor patterns of male euglossine bees evoked by floral fragrances. Animal Behavior 19:583-588.

Hills, H. G., Williams, N. H. and Dodson, C. H.   1968.   Identifi-
    cation of some orchid fragrance components.   Amer. Orch.
    Soc. Bull. 37:967-971.

Nierenberg, L.   1972.   The mechanism for the maintenance of
    species integrity in sympatrically occurring equitant Onci-
    diums in the Carribean.   Amer. Orch. Soc. Bull. 41:873-882.

Pijl, L. van der and Dodson, C. H.   1966.   Orchid flowers, their
    pollination and evolution.   University of Miami Press, Coral
    Gables, Florida.

Vogel, S.   1963.   Das sexuelle Anlochungsprinzip der Catasetinen-
    und Stanhopeen-Blüten und die wahre Funktion ihres sogenann-
    ten Futtergewebes.   Österr. Bot. Zeitschr. 110:308-337.

Williams, N. H. and Dodson, C. H.   1972.   Selective attraction of
    male euglossine bees to orchid floral fragrances and its
    importance in long distance pollen flow.   Evolution 26:84-95.

STUDIES OF NECTAR-CONSTITUTION AND POLLINATOR-PLANT COEVOLUTION*

Herbert G. Baker and Irene Baker

Botany Department
University of California
Berkeley, California

INTRODUCTION

The change in title of this paper results from the increase in scope of our on-going study of floral liquids since the program of the Congress was made up. By the same token it emphasizes that it is not an attempt to present the last word on the subject. The botanical, entomological, zoological, ecological and biochemical coverage required for a comprehensive treatment would be enormous and it is far beyond our capacity and competence. This is a status report, with suggestions for future work.

FLORAL NECTAR

The adaptive significance of nectar, produced from nectaries within a flower, is generally considered to lie in its attractiveness to potential pollinators, for whom it represents an aqueous solution of energy-providing sugars. This view, first espoused by Sprengel (1793), has held sway until the present day, although always accompanied by the spectre of an earlier belief that nectar is simply an "excretion" from the plant. The latter view, fed by observations of the existence of extra-floral nectaries as well as floral nectaries, found its strongest supporter in Bonnier (1878). Beginning with Pontedera, in 1720, proponents of this alternative view pointed out that very often the site of nectar production is the base of a superior ovary (or just above an inferior one) and, if not "stolen" by flower-visitors, the nectar may be re-absorbed into the plant. This led to the view that its function lies in the nutrition of the developing ovary and the seeds contained in it. Although resorption of residual nectar is a fact, at least in certain circumstances (Agthe, 1951; Lüttge, 1961, 1962b; Ziegler and Lüttge, 1959), we do not think we need take seed-nourishment seriously as the main function of floral nectar. Consequently, in looking at the chemical constitution of nectar we are probably justified in concentrating our attention

---

*Title in program of the Congress: "Nectar amino acids and pollinator-plant coevolution".

on adaptation to the needs and preferences of flower-visiting (anthophilous) animals that may be effective pollinators.

Despite the common view that nectar is just sugar-water, an array of other chemicals have been reported from the nectars of various flowering plants (Table 1). From this collection of nectar-constituents we have selected five groups of substances as particularly likely to be involved in different ways in co-evolution with anthophilous animals. Thus, (a) the amino acids are likely to have at least a minor nutritional influence (Baker and Baker, 1973a,b); (b) the proteins may have a similar nutritional significance or may be present as enzymes; (c) lipids may be an energy source (like sugars) or may be involved in body-building; (d) ascorbic acid (otherwise known as vitamin C) may have a nutritional function or simply act as an antioxidant in the nectar; and (e) alkaloids could have a toxic effect upon flower visitors and thereby function protectively against nectar-robbing by flower visitors that are not efficient pollinators. We have a spectrum of kinds of potential interaction and opportunities for co-evolution.

Amino acids in floral nectar

Without known exception, the watery solution that is nectar contains dissolved sugars and these are utilized by anthophilous animals for themselves and indirectly for their brood if they should be raising one. Usually it has been assumed that protein-making materials will be obtained elsewhere, from larval feeding (as in Lepidoptera), from pollen (as in many Hymenoptera, Coleoptera and Diptera) or (as in vertebrates) by the consumption of insects. However, although restricted to a liquid diet, some "higher" butterflies are known to be attracted to decaying flesh, faeces, urine and stagnant water, as well as to honey-dew and the phloem-sap which flows from wounds in plants (Ford, 1945; Klots, 1958; etc.). Some adult butterflies feed at rotting fruit (Young, 1972). Moths are known that drink fruit juices, sweat, secretions from the eyes of animals and, rarely, mammalian blood (Bänziger, 1971). Recently, Gilbert (1972) has shown that butterflies of the genus *Heliconius* collect pollen, steep it in nectar, and subsequently ingest the amino acids that diffuse from the grains.

Ziegler (1956), Lüttge (1961, 1962a), Maslowski and Mostowska (1963), Nair et al. (1964), Kartashova and Novikova (1964), and Mostowska (1965) identified amino acids in the nectars of a small number of flowering plants and, in a survey of a large number of species growing in California, Great Britain and Costa Rica, we have been able to show that nectar almost unexceptionally contains detectable amounts of amino acids (Baker and Baker, 1973a,b). In many cases these amino acids appear to be present in quantities

Table 1. Chemicals other than sugars reported to be
present in the nectars of flowering plants.

Amino acids Proteins Lipids Antioxidants (including ascorbic acid) Alkaloids	See text
Glycosides	Pryce-Jones (1944) Kozlova (1957)

Thiamin, Riboflavin, Nicotinic
  Acid, Pantothenic Acid,
  Pyridoxin, Biotin, Folic
  Acid, Mesoinositol
	Ulrich (1938)
	Haydak et al. (1942)
	Ziegler, Lüttge and Lüttge (1964)

"Organic acids" - Fumaric,
  Succinic, Malic, Oxalic,
  Citric, Tartaric,
  α-ketoglutaric, Gluconic,
  Glucuronic, etc.
	Beutler (1930)
	Niethammer (1930)
	Lüttge (1961)

Allantoin and Allantoic Acid	Lüttge (1961)
Dextrins	Rychlik and Federowska (1963)
Inorganic substances	von Planta (1886) Pryce-Jones (1940) Lüttge (1966) Waller (1973)

adequate to be of some nutritional consequence in the diets of the flower-visitors. Since these results and conclusions from them were sent to press, we have accumulated much more data on plants growing in California and in Costa Rica. All our previous conclusions have stood up to the test of these extra data (with improvement of its statistical significance in one case) and, as they are briefly recapitulated here, the larger compendium of data will be drawn upon for illustration, as well as for the substantiation of new conclusions.

Thus, at the time of writing, we have data on the concentrations of amino acids in nectars from 544 species of flowering

plants growing in California. Of these, more than 250 are California natives, the remainder being introduced species growing
with or without the benefit of cultivation. In addition, nectars
from the flowers of 169 species of flowering plants growing in
Costa Rica have been sampled. Most of these Costa Rican samples
were collected by Dr. Paul A. Opler as part of a cooperative study
of two lowland tropical forest ecosystems being carried out under
the aegis of the Organization for Tropical Studies. Results and
conclusions from the Costa Rican investigations will be referred
to occasionally but a major publication elsewhere will deal with
them in detail. Similarly, attention will scarcely be given here
to a substantial body of data that we have acquired dealing with
"extra-floral" nectar, produced by nectaries situated elsewhere
than in flowers.

In our studies, nectar was collected from each kind of flower
at the time of maximum production, during fair weather, avoiding
hot spells and any other clearly unusual environmental conditions.
Particular care was taken to avoid concentration of the nectar by
evaporation, a circumstance most likely to occur where the nectar
is openly exposed (see Pryce-Jones, 1944; Fahn, 1949; Beutler,
1953; Butler, 1954; Shuel, 1955; etc. for reviews of factors that
may influence nectar volume and at least the concentration of
sugars dissolved in it). Sugar contents (determined as "sucrose
equivalents" with the aid of a pocket refractometer), where
examined in our nectar samples, tend to be rather low (mean of 42
determinations = 27.3%) compared with figures often quoted in the
literature (cf., Beutler, 1930; Pryce-Jones, 1944; Fahn, 1949;
etc.), probably because of our use of this freshly-exuded nectar.

Each nectar collection was made by the very careful inser-
tion into the appropriate part of a flower of a very finely drawn
capillary tube, taking strict precautions to avoid any contamina-
tion of the nectar by pollen. Nectar drops placed on Whatman No.
1 chromatography paper were dried quickly and then stained with
ninhydrin (0.2% in acetone), and up to 24 hours at laboratory
temperature allowed for the usually violet color to come to a
maximum. A comparison scale for color depth was created by making
spots from dilutions of an aqueous solution of histidine and
staining these with ninhydrin (Table 2). It was found that fading
of the color with time was substantially slowed if 10% sucrose was
used instead of water in making the dilutions. The colors were
unchanged. Every sample was scored against this "histidine scale"
where each unit advance represented a doubling in concentration.
A similar scale was created with leucine solutions and gave similar color depths, but it was the "histidine scale" that was used
throughout for the comparative studies of amino acid concentrations. Some amino acids give colors other than violet with ninhydrin, most obviously proline and hydroxyproline (yellow) and

Table 2. Concentration of aqueous histidine solution required to produce depths of color scoring 0-10 in dried spots when ninhydrin solution is added.

Score on "histidine scale"	Concentration of aqueous solution of histidine
0	< 49 μM
1	49 μM (7.58 μg/ml)
2	98 μM
3	195 μM
4	391 μM
5	781 μM
6	1.56 mM
7	3.13 mM
8	6.25 mM
9	12.50 mM
10	23.00 mM (3.90 mg/ml)

asparagine (orange-brown), but on the whole it is felt that comparisons of sample spots with the "histidine scale" provide estimates of α-amino acid concentrations that are free from gross error (see Baker and Baker, 1973b for a more complete discussion of methods).

Certain micro-organisms are capable of growth in nectars, particularly those with lower sugar-concentrations. Osmophilic yeasts and some bacteria can be found in some nectars that have been exposed for a period of time (see Lodder and Krieger - van Rij, 1952; Pryce-Jones, 1957; Lüttge, 1961; etc.). However, we do not believe that these organisms have played any significant role in producing the amino acid concentrations that we have found (see discussion in Baker and Baker, 1973b), particularly as we have utilized freshly-secreted nectar consistently.

A weak positive correlation (r = +0.37) between sugar content of nectar samples and their "histidine-scale" scores (assuming a linear relationship) was noted previously (Baker and Baker, 1973b). This has been sustained with the accumulation of further data (viz. r = +0.36, using 42 samples). Because it takes at least 15 microliters of nectar to be able to use a pocket refractometer to measure sugar content, these results are necessarily based on the species that are more prolific nectar-producers. However, there is no apparent correlation between the amount of nectar produced and its "histidine scale" score (Baker and Baker, 1973a,b), so the results of the sugar/amino acid comparison need not be considered unreliable on this basis.

In our previously published work (Baker and Baker, 1973a,b) we showed that although amino acids are present in the nectars of plants at all levels of "advancement" on such evolutionary scales as those put forward by Sporne (1969) and other phylogenists, there is some evidence of increase in concentration and in consistency of concentration from plants showing a character generally accepted as "primitive" to others showing the "advanced" state of this character (Table 3). The point to be made in this symposium is that, in general, "advancement" of this sort parallels "advancement" in pollinator-relationships (with increasing special-ization to flower-visitors possessing greater discriminatory powers, greater ability to reach concealed nectar and greater flower-specificity).

The contrast between woody and herbaceous plants in Table 3 has been examined further, with all the data available to date. Table 4 shows that the mean "histidine-scale" score for herbaceous plants is significantly higher than that for all woody plants. In the latter group, the vines give a high mean score, probably largely because of the prominent representation of the high-scoring Leguminosae and Convolvulaceae in this small sub-group. Trees, and especially shrubs, give a low average compared with the herbs and this has particular interest when taken in conjunction with the investigations of sugar contents of nectars by Percival (1961).

Percival (1961)   roughly divided the 889 species whose nectar she examined into three groups (which we may call I, II, and III). Group I consists of nectars in which sucrose predomi-nates over glucose and fructose; in Group III the situation is reversed. In Group II there is a more or less equal balance. In presenting her results, Percival keeps Hutchinson's (1959) basically woody line Lignosae separate from the basically herba-ceae. No opinion is expressed here as to the validity of this separation phylogenetically but it has its value for our limited comparative purposes. One hundred and thirteen of the species in our study are also represented in Percival's list and another 71 are closely related members of genera of families for which there

Table 3. Comparisons of mean "histidine scale" scores for species grouped according to whether they show "primitive" or "advanced" states of a particular character.

	No. of spp.	H.S. mean	c.v.	P
Woody	92	4.45	48%	<< 0.001
Herbaceous	174	6.00	32%	
Open nectar	73	4.88	48%	≃ 0.01
Concealed nectar	193	5.66	36%	
Polypetalous	105	4.75	45%	<< 0.001
Sympetalous	161	5.90	34%	
Actinomorphic	158	5.15	42%	≃ 0.001
Bilateral	108	5.99	34%	
Many stamens	47	4.51	51%	< 0.005
Few stamens	219	5.66	37%	
Polycarpous	24	4.88	39%	≃ 0.10
Syncarpous	292	5.54	39%	
Hypo- and peri-gynous	168	5.29	42%	< 0.60
Epigynous	98	5.46	37%	

is enough evidence about sugar balance for determinations to be made by reasonable inference. Of these 184 cases, only seven fall into Group II and these are not considered further. Table 5 shows the results with Groups I and III.

In considering these results, we should keep in mind Percival's deduction from her survey that there is a tendency for flowers with deeply concealed nectar to produce nectar classifiable in Group I, i.e., with sucrose predominant. The evolution of "sophisticated" pollination mechanisms has gone farther in the Herbaceae than in the Lignosae and it is striking that the mean "histidine scale" score for Herbaceae is significantly higher than that for the Lignosae. It is also notable that there is no significant difference between the mean "histidine scale" scores for Group I (sucrose predominance) and Group III (fructose and glucose predominance) in the Lignosae but that there is a very

Table 4. Comparisons of mean "histidine scale" scores for species grouped according to life form.

	No. of species examined	Mean H.S. score	Coeff. of variation	P
All determinations	544	5.05	39%	
Herbs	362	5.41	33%	$\ll 0.001$
Shrubs	125	4.16	52%	$0.70 < P < 0.80$
Trees	39	4.30	47%	$0.70 < P < 0.80$
Vines	18	5.61	32%	$\ll 0.001$
All woody plants	182	4.33	49%	

Table 5. Comparisons of mean "histidine scale" scores for species grouped according to whether disaccharides or monosaccharides predominate in the nectar, and on the basis of classification in Hutchinson's "phylogenetic" lines Lignosae and Herbaceae.

	No. of species	Mean H.S. score	c.v.	P
Sucrose dominant (Group I)	103	5.79	31%	
Fructose + glucose dominant (Group III)	74	4.89	34%	< 0.001

Lignosae

	No. of species	Mean H.S. score	c.v.	
Group I	31	4.63	46%	<< 0.001
	P = < 0.90			
Group III	28	4.73	40%	0.50

Herbaceae

	No. of species	Mean H.S. score	c.v.
Group I	72	6.28	22%
	P = << 0.001		
Group III	46	4.99	30%

highly significant difference in the Herbaceae.

These suggestions of a strong connection between the evolution of pollination mechanisms and the amino acid concentrations in nectar fit well with our analysis on the basis of particular types of pollination systems. Table 6 shows the mean "histidine scale" scores for the plants that are visited by (and, apparently pollinated by) flower-visitors of various sorts.

Most flowers are visited by more than one kind of anthophilous animal. In presenting the data here, decisions were made as to the most important class of visitors in each case. For this, the literature available was consulted (especially, Müller, 1883; Knuth, 1906-9; Kirchner, 1911; Robertson, 1928; Meeuse, 1961; Free, 1970; Faegri and van der Pijl, 1971; and Proctor and Yeo, 1973) as well as our own experience and appreciation of the structure and behavior of flowers. For 378 species, one class of pollinator is clearly predominant, but for the remainder no simple ascription is possible. Eighty-five species fall into a special category of species that are visited freely both by butterflies and bees. Seventy-three other species are visited by two classes of pollinator and are considered in relation to each one (mostly this concerns short-tongued bees and flies visiting flowers with openly displayed nectar). Eight species represent the extreme of having three classes of pollinator without clear predominance of any one.

The first notable contrast is that between the mean "histidine scale" scores for flowers adapted to short-tongued bees (and other short-tongued visitors) and those adapted to long-tongued bees (with probosces more than 6 mm in length). This difference could be seen in our previous results (Baker and Baker, 1973a,b) but was not statistically significant. Now, with extra data available it is very highly significant.

The long-tongued bees visit flowers that have greater depth than those available to the short-tongued bees (and we have seen from Table 3 that flowers with concealed nectar give higher scores on the average than those with exposed nectar). Similarly, the "long-tongued bee flowers" are more frequently bilateral (zygomorphic) and such plants score higher, on the average, than actinomorphic ones do (Table 3). [Incidentally, in Bavaria, in two seasons, Leppik (1953) recorded 2,756 bumblebees and only 22 honeybees visiting bilateral flowers whereas 2,779 honeybees and only 324 bumblebees visited actinomorphic flowers].

But these correlations are not explanations for the differences in nectar amino acid contents. Perhaps one explanation may be suggested by what is known of larval feeding in bumblebees. Free and Butler (1959, p. 15) reviewed the subject and concluded from the work of Bailey (1954) that bumblebees can only digest relatively small quantities of pollen, at least by comparison with

Table 6. Comparisons of mean "histidine scale" scores
for species grouped according to their
adaptations to types of pollinator.

Pollinator type of flower	No. of species examined	Mean H.S. score	Coeff. of variation	P
Short-tongued Bee	208	4.59	41%	} << 0.001
Long-tongued Bee	96	5.64	29%	
Bee + Butterfly	86	5.49	26%	} 0.01
Butterfly	40	6.40	24%	
Settling Moth	14	5.75	31%	} 0.10 < P < 0.20
Hovering Moth	25	4.86	48%	
Generalized Fly	68	4.35	45%	} << 0.001
Specialized Fly	9	9.00	(14%)	
Beetle	7	7.86	18%	
Wasp	6	5.58	40%	
Old World Bird	21	3.31	47%	
Hummingbird	49	4.86	46%	
Bat	4	3.75	61%	

honeybees. And the bilateral flowers that they visit usually produce less pollen than the wide-open actinomorphic flowers with many stamens. If pollen contributes less to the nutrition of these bigger bees, an increased amino acid intake from the nectar could be beneficial and selectable. However, in considering this possibility, we should remember that bumble bees may not be representative of all long-tongued bees — and honeybees are certainly not typical of short-tongued bees.

The rather large group of flower-species that are visited frequently by both butterflies and bees (usually long-tongued), including many Compositae, shows a mean "histidine scale" score comparable to that of the long-tongued bee flowers, but this is significantly lower than the mean score for clearly-recognizable "butterfly" flowers. As pointed out elsewhere (Baker and Baker, 1973a,b), protein-building materials for adult Lepidoptera are traditionally thought of as being acquired during a phytophagous (or, more rarely, carnivorous) larval life and carried over to the adult in the fat body. For butterflies and moths (such as many Satyridae and Saturniidae) which do not feed as adults, larval feeding must be the sole source of nitrogen for life-prolongation and the facilitation of reproduction - and it may be significant that the adult lives of these Lepidoptera are over in a few days. At the other extreme, large tropical forest butterflies like *Morpho* spp., which feed as adults on rotting fruits of trees such as the legume *Coumarouna (Dipteryx) oleifera* (from which they must get more than just sugars and water) have life-spans of several months (Young, 1972). In addition, Gilbert (1972) has pointed to the extended life-spans and high reproductive outputs of the adults of *Heliconius* which get amino acids from pollen which they douse in nectar.

Our results (Table 6) show that even those butterflies which merely collect nectar can also acquire amino acids as adults. Judging by their mean score on the "histidine scale" (i.e., 6.40), about 0.4 ml of "butterfly flower" nectars would contain about 840 nmols· of amino acids (and a daily intake of this amount is believed by Gilbert (1972)   to have a profound effect on the life-span and reproductive output of *Heliconius*). It takes no more than 20 flowers of *Dianthus barbatus*, a typical "butterfly flower", to provide such an amount of nectar, and smaller quanti- ties than this are likely to have some effect, even if only as phagostimulants. Zebe (1953) has reported that butterflies which he observed drank a quarter of their own weight in 40% sugar solu- tion, meaning that at one sitting they would absorb about 0.05 g of nectar (containing about 100 nmols of amino acids).

Even so, it must be noted that some very characteristic "butterfly" flowers give only moderate "histidine scale" scores when their nectar is collected very carefully. This is the case

with such species as *Buddleia davidii* (Loganiaceae) (scores = 5-6),
Syringa vulgaris (Oleaceae) (score = 4), *Leptodactylon californicum* (Polemoniaceae) (score = 4), as well as several Apocynaceae.
In each of these cases, the structure of the flower, with
introrsely dehiscing anthers contained within the narrow corolla-
tube, is such that the first butterfly proboscis to penetrate to
the nectar at its base will inevitably push pollen down into that
nectar where its amino acids will be added to those already present. Subsequent butterfly visitors, even those coming only a few
minutes later, may then suck up a nectar that is much richer in
amino acids. Experiments in which fine glass capillaries were
used to simulate butterfly probosces raised the histidine scale
score of *Leptodactylon californicum* from 4 to 7 in only one minute. Higher-scoring "butterfly flower" nectars, like those of
Duranta repens and species of *Heliotropium*, may be raised to even
greater heights in the same way.

Also, uncollected nectar accumulates in the corolla-tubes of
such flowers as those of *Jasminum officinale* and may reach the
level of the anthers and the pollen in them. When this happens,
the "histidine scale" score of the nectar is raised (in our
observations from ca. 4 to ca. 7). The same enrichment may take
place in flowers of "bee + butterfly" Compositae.

Very similar to butterflies in their flower-visiting behavior
are those moths which do not hover but regularly settle on or in
the flowers they visit. Belonging to such families as the Noctuidae, Zygaenidae and Pyralidae, these moths contrast in behavior with
the swift-flying, hovering hawkmoths (Sphingidae). The flowers
that play host to the "settling" moths have nectar which is rather
rich in amino acids (and may be fortified by contact with pollen
as in "butterfly flowers") but the mean score is rather lower than
that for "butterfly flowers". Their nectar flow tends to take
place nocturnally rather than diurnally and this may influence its
amino acid content. Indeed, all nocturnally-produced nectars may
be expected to be rather poor in amino acids if their composition
reflects that of the phloem sap in the plants that produce them,
for Ziegler (1956) has demonstrated that in the sap of *Fraxinus
americana* ninhydrin-positive substances are absent at night, re-
appear at about 0700 hours and reach their maximum concentration
in the late morning or early afternoon. In fact, a series of
experiments that we have carried out with *Jasminum officinale*, which
secretes nectar by day and by night, has shown that nocturnally-
produced nectar is noticeably weaker than that produced during the
day (Table 7). Nevertheless, Zebe (1954) showed that hungry
individuals of the "settling" moth *Agrotis pronuba* (Noctuidae) will
drink 150 mg of nectar at one time, which, in our calculations,
would contain about 164 nmols of amino acids.

"Hawkmoth flowers" show the lowered mean "histidine scale"

TABLE 7. Mean "histidine scale" scores of
nectar collected from flowers of
Jasminum officinale at different
times of day and night.

Time	No. of observations	Mean H.S. scale score
07.00-07.20	24	4.0
13.00	10	4.3
15.45	4	4.1
19.00-20.00	16	3.4
23.20-23.30	12	3.5

score of nocturnally visited flowers to still greater degree
(Table 6). The hawkmoths need, and get, large quantities of nectar from the flowers they visit on their energy-expensive darting
flights (Heinrich, 1971). Possibly, if the amino acid-to-sugar
ratios of these nectars were comparable with those of the nectars
in "settling moth" and "butterfly" flowers, amino acid toxicity
might result (see discussion in House, 1965, p. 776-7). A feeding
experiment with *Manduca sexta* showed that it may drink more than
a milliliter of nectar substitute (5% honey water) without being
satiated (B. Hammock, L. Hammock and R. Dietz, unpublished).
    Fragmentary data from "bat flowers" obtained in California
and the considerably greater amounts of data becoming available
from Costa Rica for such flowers indicate that, there, the nocturnally produced nectars are also relatively weak in amino acids.
    There is a striking difference between the overall weakness
(though considerable variability) of nectars prominently displayed in "generalized fly flowers" and the considerable amino
acid strength of the nectars (when produced) of the highly specialized flowers that lure carrion- and dung-flies away from their
normal nitrogen-rich food and egg-laying haunts. This is fully
discussed in Baker and Baker (1973a,b). Only one qualification of
the picture is needed here: the secretions available to flies in
flowers of *Aristolochia* are properly described as stigmatic
exudates rather than as nectar. However, they seem to play a
similar role in nourishing the visitors (Baker, Baker and Opler,
1974).
    The special case of flowers visited by so-called "biting

flies" will not be dealt with here except to say that these may be
very important in the Arctic, where the researches of Hocking
(1953, 1968 and 1971) and his students have shown that nectars of
high sugar concentration are produced. If the partial correlation
between sugar concentration and amino acid concentration that we
have found to be generally applicable should apply in this particular case, it may be expected that the amino acid concentrations
of "biting fly" nectars will also be above average. A number of
the flower species listed by Hocking (1953, 1968), Sandholm and
Price (1962) and Thien (1969) as being visited by mosquitoes in
northern latitudes have been examined and their "histidine scale"
scores are generally high.

"Bird flowers" constitute another special problem. There is
evidence (Baker, 1973, etc.) that relationships between nectar-
drinking birds and nectar-producing flowers have evolved several
times independently in different parts of the world. In each case,
however, it appears that some other means of food acquisition
(almost always the consumption of insects) has been retained, and
that, for the birds, nectar may be chiefly liquid refreshment and
a readily utilizable energy source. Consequently, it would not be
surprising if the "histidine scale" scores for the nectars of
"bird flowers" should be rather low, for they could never rival
insects as nitrogen sources. For the "Old World bird flowers"
this is strikingly true; for the flowers from the New World that
are visited by hummingbirds a somewhat different picture is presented. Most of the "hummingbird flowers" investigated are from
species that grow in western North America, and most of these
belong to genera in which the majority of the plants are
pollinated by other kinds of visitors - usually insects. Consequently, it seems that they still retain relatively high concentrations of amino acids from their ancestors.

Grant and Grant (1968, p. 94) have pointed out that "A highly
developed bee pollination system was evidently the breeding ground
of many new lines of bird-pollinated plants in western North
America". Presumably, these were long-tongued bees, whose
flowers have nectars relatively rich in amino acids. Grant and
Grant contrast this relatively recent evolution in temperate
regions with the situation relating to "the more ancient hummingbird flowers in the American tropics," and, in this connection, we
may point out that our data from Costa Rica substantiate this
fully, for the "histidine scale" scores for "hummingbird flowers"
collected there are among the lowest of all the various sorts.

Very few "beetle flowers" and "wasp flowers" were available
to us in California. Consequently, discussion of their nectar
amino acids will be deferred until the much larger samples from
Costa Rica are treated elsewhere.

At the time of writing, paper chromatography analyses of the nectars of 60 flower species have been carried out so as to reveal which amino acids are present. Table 8 presents a summary of the

Table 8.   Summary table of amino acids identified in samples of nectar from plants growing in Berkeley. For each listed amino acid the number of occurrences is given for each pollinator type category and should be considered in proportion to the number of species samples for each of these categories.

		BEE	BUTTERFLY	MOTH	FLY	BIRD	OTHERS
	Number of species	29	25	9	6	11	2
"Essential"	Arginine	3	4	2		2	1
	Histidine	11	12	2	3	4	1
	Lysine	10	12	1	2	2	1
	Tryptophan		1		2		
	Phenylalanine	2	2		1	1	
	Methionine	4	2			1	
	Threonine	8	4	2	2	4	1
	Leucine/Isoleucine	8	6	2	4	2	
	Valine	6	6	2	4	1	
"Quasi-E"	Serine	19	17	5	6	7	1
	Glycine	13	13	4	6	5	1
	Proline	3	2		1	1	
Common	Alanine	9	11	2	6	4	
	Aspartic acid	18	16	6	2	5	1
	Glutamic acid	14	16	4	4	7	1
	Others	13	16	7	6	5	5

findings (which are qualitative, not quantitative). There is
great variability from species to species and the number of amino
acids identified by this means varies, in individual cases, from
one to twelve. More refined methods of analysis might increase
the numbers of amino acids in particular cases, but the extra
acids would be present in very small quantity.

As pointed out elsewhere (Baker and Baker, 1973a,b) all ten
of the amino acids usually regarded as essential in insect
nutrition (Haydak, 1970; Dadd, 1973) are present in the nectar of
one species or another, as are the three "quasi-essential" acids.
In our sample of flowers pollinated by butterflies (for whom the
nectar amino acid balance may be of great importance) the roster
of "essentials" and "quasi-essentials" is complete (Table 8).
Overall, serine, glycine, alanine, aspartic acid and glutamic acid
are the most frequently available while tryptophan, phenylalanine
and methionine are the "essentials" hardest to come by. Proline,
which is so abundant in many pollens (Auclair and Jamieson, 1948;
Linskens and Schrauwen, 1969; Echigo, 1971; etc.) is also rather
infrequent in nectar. It is relatively abundant in the nectar of
Callistemon sp. and *Melaleuca ericifolia* (which represent two
closely related genera of the Myrtaceae - a family which we are
investigating more closely for this reason and for the interesting
relationships with anthophilous animals that it has developed,
particularly in Australia).

The historical context in which to place our analyses of
individual acids is not large. Probably the first identification
of individual amino acids in nectar was made by Lüttge (1961) who
found alanine, threonine and glutamic acid in the extra-floral
nectar of *Sansevieria zeylanica* and glutamic acid, aspartic acid,
methionine, serine, tyrosine, cystine, proline and alanine (in
descending order of concentration) in the nectar of *Musa sapientum*. Maslowski and Mostowska (1963), Nair *et al.* (1964), Mostowska
(1965), and Kartashova and Novikova (1966) identified amino acids
in the nectars of a number of species, with as many as 13 different amino acids represented in the same nectar. Generally
speaking, in the reported cases, aspartic acid, glutamic acid,
serine, glycine and alanine appear to be the commonest nectar
constituents, which agrees with our findings (Table 8). It should
be emphasized, however, that as yet nothing is known as to the
constancy or variability of the presence of particular amino
acids within a species in the face of genetical and environmental
variation. This, and the degree of differentiation between species
of the same genus, is an important subject for our future investigations.

Proteins in floral nectar

In 1936, Beutler and Wahl reported the presence of the enzyme invertase in floral nectar from a species of *Tilia* and, later, Zimmerman (1953, 1954) recorded transglucosidase in nectar from flowers of *Robinia pseudacacia* and transfructosidase in extra-floral nectar from *Impatiens holstii*. Phosphatases have been reported by Cotti (1963) and Zalewski (1966) in the nectars of a number of species while Zauralov (1969) has recorded the presence of oxidases. Tyrosinase occurs in the nectar of *Lathraea* according to Lüttge (1961).

More substantial quantities of proteins that are not necessarily enzymic in nature, have been reported from (or implied to be present in) floral nectar by a number of investigators (e.g., Buxbaum, 1927; Ewert, 1932; Pryce-Jones, 1944; Lüttge, 1961). Baker and Baker (1973b) reported the presence of protein in the nectars of *Calluna vulgaris, Erica mediterranea* and *Bergenia crassifolia*. Honeys from *Calluna vulgaris* and *Fagopyrun* sp. are famous for their high protein content (Pryce-Jones, 1950).

In the older literature there is often the implication that the presence of proteins in nectar is due to microbial action. Bacteria and, particularly, yeasts are known that can flourish in nectar (Pryce-Jones, 1950; Lodder and Krieger-van Rij, 1952; Lüttge, 1961; etc.). If the nectar lasts for several days in a flower (as it does with *Musa sapientum*, studied by Lüttge (1961)who investigated the contents of fallen flowers in a green-house in Germany), the amount may become sufficient (in this case 2.75 to 5.85 mg per ml of nectar) to be of nutritional value to a flower-visitor. However, in most cases, it will be relatively freshly secreted nectar that is consumed by anthophilous visitors and the question arises as to whether this ever contains proteins in significant quantity.

In our studies dried nectar spots were tested for proteins by the application of 0.1% brom-phenol blue in methanol for 1 hour, followed by a rinse for 1/4 to 1/2 hour in 5% acetic acid. The papers were then dried and a blue color indicated the presence of protein.

Up to the present, the nectars from 139 species of flowers have been tested in this manner. Table 9 shows the composition of this sample in terms of flower visitors, in comparison with that for the ninhydrin tests (which represent the total sample of species used in our studies). It will be seen that, although considerably smaller in total number of species, the sample used for protein-testing is a reasonably fair one. Therefore, it is probably meaningful that only 19 out of 139 (14%) of the species gave positive results (see Appendix I). In many of these cases, the positive result would have to be placed at the "trace level.

Table 9. Percentages of the various pollinator classes in samples of total species available which were *examined* for the various chemicals in their nectars.

Examined for:	Amino acids	Lipids	Antioxidants	Alkaloids	Proteins
Number of species	544	220	210	126	139
Mean "histidine scale" score	5.05	5.06	5.06	5.08	5.24
% short-tongued bee	38.2	40.0	39.0	35.7	34.5
% long-tongued bee	17.5	20.9	21.4	19.0	16.5
% bee + butterfly	15.6	13.6	13.3	11.1	9.4
% butterfly	7.4	9.5	10.0	11.1	8.6
% settling moth	2.6	2.3	2.9	4.0	2.2
% hovering moth	4.8	8.2	7.6	10.3	7.2
% generalized fly	12.5	15.9	14.8	11.1	6.5
% specialized fly	1.7	0.5	0.5	0.8	2.2
% wasp	1.1	0.9	1.0	1.6	1.4
% beetle	1.3	2.7	2.9	1.6	1.4
% hummingbird	9.0	7.3	12.9	15.1	13.7
% old world bird	3.9	3.2	3.3	1.6	2.7
% bat	0.7	0.9	1.0	1.6	1.3

"Histidine scale" scores for the species giving positive results
are not significantly different from those for species giving
negative results (Table 10), so it does not appear that nectar-
proteins are merely broken down into amino acids in those cases
where negative results for protein are obtained.

No clear taxonomic picture emerges from the lists of "positives" and "negatives" although there is a suspicion that "more
advanced" taxa are less well represented among the "positives"
than "less advanced" ones.

Table 10 also shows a breakdown of the results by pollinator-
type (the total number of scores being 150 rather than 139 because
some of the flower species are frequently associated with more
than one kind of flower visitor). The data are arranged in
descending order of the ratio "positive" to "negative". Even
ignoring the results for categories in which very few determinations were made, it appears that organisms whose digestive
systems might be expected to be able to cope with proteins as such
are in the upper part of the table, whereas the Lepidoptera, whose
guts do not produce appropriate enzymes, as far as is known
(Wigglesworth, 1966), are low down. However, in this connection,
the lowly positions of "bird flowers" are anomalous.

In general, the pollinator classes of flowers showing the
higher positive to negative ratios are associated with exposed or
only slightly concealed nectar; where the nectar is deeply hidden
(as with "long-tongued bee", "lepidopteran" and "bird" flowers),
the proportion is very low. This was examined further by taking
advantage of the fact that 65 of the species examined for proteins
are either the same as or are closely related to species whose
nectar-sugars were examined by Percival (1961). Consequently, the
balance between sucrose and glucose plus fructose can be stated
fairly reliably for these species. Six out of 23 of the species
which have glucose plus fructose predominating over sucrose
(Group III) gave positive results for protein. Only 4 out of 39
of the species in which sucrose predominates (Group I) gave positive results. Three species with balanced sugar proportions
(Group II) gave negative protein results. The difference between
the proportions for Groups I and III is almost significant
statistically (χ^2 = 2.34; 0.1 > P > 0.05). Because Group I flowers
tend to be those which have concealed nectar (Percival, 1961), this
fits with the suggestion made above that "more advanced" flowers
are less likely to contain protein in the nectar. The results
would also be consonant with a suggestion that nectars which contain a predominance of monosaccharides have undergone enzymatic
attack - and they are the ones with the greater tendency to contain detectable protein.

Consequently, although we have also seen some correlation
between protein presence and the ability of a pollinator to digest

Table 10. Mean "histidine scale" scores of nectar samples
giving positive and negative results for
proteins, together with an analysis of
"positives" and "negatives" grouped according
to pollinator type.

Mean H.S. score of "positives" 5.48 ⎫
Mean H.S. score of "negatives" 5.19 ⎬ 0.50 > P > 0.40
Mean H.S. score of all tested 5.25 ⎭

	Positives(P)	Negatives(N)	Ratio $\frac{P}{N}$
Bat	1	1	1.00
Specialized Fly	1	2	0.50
Bee + Butterfly	3	10	0.30
Generalized Fly	2	7	0.29
Short-tongued Bee	9	39	0.23
Long-tongued Bee	3	20	0.15
Hovering Moth	1	9	0.11
Butterfly	1	11	0.09
Hummingbird	1	18	0.06
Old World Bird	0	4	0.00
Settling Moth	0	3	0.00
Beetle	0	2	0.00
Wasp	0	2	0.00

proteins, the lightness of the color depths in the positive tests, together with the sugar-balance evidence, suggests that it may be as enzymes, rather than as foodstuffs, that the proteins have their biological significance in nectar.

Lipids in floral nectar

The presence of lipids in some nectars seems to have gone un-noticed until it was reported by us recently (Baker and Baker, 1973b). However, Voge (1969) has described the occurrence of oil drops as an *alternative* to nectar in the flowers of some bee-pollinated members of the Scrophulariaceae, Malpighiaceae and Orchidaceae (secreted by external glands that he calls "elaio-phors") (see also Faegri and van der Pijl, 1971, p. 79).

In our studies, lipids in nectar were tested for by making spots in the usual manner and then applying a 1% aqueous solution of osmium tetroxide. Lipids derived, at least in part, from unsaturated fatty acids give a black color with this test (see discussion in Cain, 1950, for the limits of reliability of this test). They were also tested for by staining the oil globules (visible in nectars placed on glass slides) with Sudan III or Nile Blue.

The nectars of *Jacaranda ovalifolia* (Bignoniaceae) and *Trichocereus andalgalensis* (Cactaceae) are voluminous enough to display a milky consistency to the naked eye, but these are only two among the 75 species out of 220 tested (34%) that gave positive results in the lipid test (Appendix II). Forty plant families are represented among the species whose nectars contain lipoidal substances. There is no obvious taxonomic pattern to the distribution of positive and negative scores. Herbs, shrubs and trees are represented among the "positives" and "negatives".

Again, Table 9 shows that the species examined represent a fair sample of the total species available to us. Table 11 de-tails the results by pollinator class, again arranged in descend-ing order of "positive" to "negative" ratio for each pollinator class.

Although in each well-represented category more species lack nectar-lipids than show them, the emphasis among the "positives" is on the Hymenoptera and Diptera, which have digestive systems that include lipases capable of breaking any lipids down into assimilable substances (Wigglesworth, 1966). Where birds and Lepidoptera are concerned, the flower species that serve the very actively flying hummingbirds and hovering (hawk) moths are involved to a greater extent than those which cater to the needs of the settling moths, butterflies and Old World birds that perch while feeding. Thus, it is tempting to see nectar-lipids as a minor energy source for some anthophilous animals, especially as

Table 11.   Mean "histidine scale" scores of nectar samples
            giving positive and negative results for lipids,
            together with an analysis of "positives" and
            "negatives" grouped according to pollinator type.

Mean H.S. score of "positives" 5.67 ⎫
Mean H.S. score of "negatives" 4.68 ⎬ P < 0.001
Mean H.S. score of all tested 5.06 ⎭

	Positives (P)	Negatives (N)	Ratio $\frac{P}{N}$
Wasp	2	0	∞
Specialized Fly	1	0	∞
Bee + Butterfly	13	17	0.76
Hovering Moth	7	11	0.64
Beetle	2	4	0.50
Short-tongued Bee	29	59	0.49
Long-tongued Bee	15	31	0.48
Generalized Fly	11	24	0.46
Hummingbird	7	19	0.37
Settling Moth	1	4	0.25
Old World Bird	2	5	0.20
Butterfly	2	19	0.11
Bat	0	2	0.00

many insects, including Lepidoptera, are known to use lipids in
their muscles as the major energy source for flight (Kozantshikov,
1938; Zebe, 1954, 1959).

However, the inclusion of some flowers visited by Lepidoptera among the "positives" is difficult to explain adaptationally
in the absence of evidence that these insects can digest lipoidal
substances (and we are in the process of investigating this possibility). It should be remembered that some insects can absorb
unhydrolyzed fats as long as free fatty acids are also present to
emulsify them (cf. Gilmour, 1965, p. 81). However, in the cases
of the moths, only traces of lipids were responsible for several
of the positive scores; in the cases of the butterflies all of
the flower species involved are also visited by bees or by birds
and the significance of their lipids may be found in relation to
these other visitors. We should remember the observations of
Vogel (1969) and his oil-collecting tropical bees.

Table 11 also shows that the ninhydrin ("histidine scale")
scores of the nectars that give positive lipid results are very
significantly greater than those giving negative results for
lipids. This matter will be discussed in more detail in connection with antioxidant presence in nectar, which is the next subject for treatment.

Antioxidants in floral nectar

Vitamin C (ascorbic acid) has been identified in commercial
honey at various times (e.g., Kask, 1938; Griebel and Hess, 1939,
1940; Haydak et al., 1942; Pryce-Jones, 1950). In nectar,
Griebel and Hess (1940) found over 200 mg of ascorbic acid per
100 ml of nectar in three species of Labiatae. Weber (1942)
identified it in nectar of *Fritillaria* and in *Impatiens* (Weber,
1951). Bukatsch and Wildner (1956) found substantial quantities
of ascorbic acid in the nectars of a number of species, particularly members of the genera *Aquilegia*, *Oenothera*, *Eichhornia*,
Impatiens, *Datura*, and *Lilium*, as well as in the extra-floral
nectar of *Centaurea*. Ziegler, Lüttge and Lüttge (1964) added six
more species. However, in these accounts, it is not certain that
the reducing agent claimed as ascorbic acid is always that substance, for aldehydes, ketones and various organic acids can
exert such action.

Consequently, our systematic program of testing nectar spots
is to be considered an examination for the presence of antioxidants including ascorbic acid. The reagent used was 2-6
dichlorophenol-indophenol (a 0.1% ethanolic solution of the sodium salt). Rapid bleaching of the red color that develops
initially (to a white or whitish-pink color) indicates the presence of reducing agents (Stahl, 1969). As a further check,

positive results were confirmed by treating spots with 0.05%
aqueous potassium permanganate to observe bleaching (Stahl, 1969).

Table 9 shows that the sample of 210 species tested is a
fair one. Sixty-three (i.e. 30%) of these nectars gave a positive result. Once again, there is no obvious taxonomic connection among the flowers whose nectars give positive results
(Appendix III). Thirty-six families are involved in the samples
and no family with more than three species included showed all
positive results. Griebel and Hess (1940) found the nectar of
Thymus serpyllum, *Mentha arvensis* and *Lycopus europaeus* (all
Labiatae) to contain plentiful ascorbic acid (*Mentha aquatica* was
the richest in the tests by Ziegler, Lüttge and Lüttge, 1964),
and we can confirm this for *Thymus*. However, whereas *Lavandula
spicata* and *Salvia mellifera* were also "positive," *Prunella vulgaris* and *Stachys bullata* gave negative results.

Table 12 lists the "positive" to "negative" ratios by pollinator type. Here, again, it is clear that the correlation is
with "histidine scale" score. The pollination classes that we
associate with high amino acid concentrations are in the upper
part of the table (e.g. "specialized fly" flowers, "butterfly"
flowers, "bee and butterfly" flowers) while near the bottom are
"bird" flowers, "generalized fly" flowers and "hovering moth"
flowers. Calculation of a rank correlation coefficient linking
this ratio with "histidine scale" score (from Table 6) gives a
high result (r_{rank} = +0.75). It was noted when lipids were being
considered that the mean "histidine scale" score for the "positives" was very significantly higher than for the "negatives."
Table 13 shows that the mean is even higher for those nectars
that contain antioxidants and highest of all for those that contain both lipids and antioxidants.

In our sample of nectars examined for both lipids and antioxidants (210 species), 43 produced only lipids, 29 produced only
antioxidants and 38 produced both (a rather higher number than
would have been expected on chance alone — χ^2 = 5.2; P > .02 —
although not significantly so). It will be reported elsewhere
(Baker, Baker and Opler, 1974) that lipids and antioxidants
appear to occur together constantly in stigmatic exudates, and it
may be that prevention of oxidation of the lipoidal substances is
one function of the antioxidants. This is further backed by the
identification of ascorbic acid in the oily elaiosomes of three
species of flowering plant that have ant-distributed seeds
(Bukatsch and Wildner, 1956).

The antioxidant is often ascorbic acid (as shown by chroma-
tograms that we have run with ascorbic acid standards for comparison) as, for example, with the nectars of *Thymus serpyllum*, *Sedum
dendroideum*, *Armeria maritima*, *Oxalis rubra* and *Limnanthes doug-
lassii* var. *sulphurea*. However, other easily oxidizable

Table 12. Mean "histidine scale" scores of nectar samples
          giving positive and negative results for anti-
          oxidants, together with an analysis of
          "positives" and "negatives" grouped according
          to pollinator type.

Mean H.S. score of "positives" 5.93 ⎱
Mean H.S. score of "negatives" 4.65 ⎰ P < 0.001
Mean H.S. score of all tested 5.06

	Positives (P)	Negatives (N)	Ratio $\frac{P}{N}$
Specialized Fly	1	0	∞
Wasp	1	1	1.00
Butterfly	10	11	0.91
Beetle	2	4	0.50
Bee + Butterfly	9	19	0.47
Short-tongued Bee	26	56	0.46
Long-tongued Bee	14	31	0.45
Hummingbird	8	19	0.42
Settling Moth	1	5	0.20
Old World Bird	1	6	0.17
Generalized Fly	4	27	0.15
Hovering Moth	2	14	0.14
Bat	0	2	0.00

Table 13. Comparisons of mean "histidine scale" scores for
nectars grouped according to the possession of
lipids and/or antioxidants.

	No. of species	Mean H.S. scale score	c.v.	P		
Lipids + antioxidants	38	6.28	27%	} 0.35	} < 0.10	} < .0(
All with antioxidants	67	5.93	31%			
All with lipids	85	5.67	31%			
Neither	100	4.43	39%			
All species tested for antioxidants and lipids	210	5.06	37%			
All species available	544	5.05	39%			

substances are often present, with or without ascorbic acid.
This is the case with *Nigella damascena, Catalpa bignonioides,
Lavandula spicata, Wisteria sinensis, Rhododendron* sp., and
Aesculus californica. It is notable that Becker and Kardos
(1939) already concluded, from animal tests, that the antioxidants in honey are not always ascorbic acid. Gontarski (1948)
reported that pharyngeal gland secretions in the honey bee contain an oxidizing enzyme that soon destroys ascorbic acid and an
ascorbic acid oxidase may even be present in nectar, itself,
according to Zauralov (1969).

     These evidences that the chemicals for which the tests have
been made may be important for their ability to prevent oxidation,
and that honey bees have the enzymic means for destroying ascorbic acid, add up to give an impression that the role of the antioxidants in nectar is not the nutritional one conjured up by the
name Vitamin C. In any case, according to Gilmour (1961), ascorbic acid is not required in the diet of most insects whose
requirements have been studied; the exceptions seem to be those
which feed on plant parts much richer than nectar in this substance (for such insects appear to have lost the ability to synthesize it).

Unfavorable substances in floral nectar

     Waller (1973) has shown that the attractiveness of the nectar
of *Allium cepa* to honey bees varies inversely with the amount of
potassium present, and even only 1,500 parts per million in sugar
solution had a deterrent effect on these insects. However, nectars that are actually *toxic* to anthophilous insects also exist
(see review in Pryce-Jones, 1944; also Mauritzio, 1945; Kozlova,
1957; Carey *et al.*, 1959; Jaeger, 1969; Bell, 1971; Palmer-Jones
and Line, 1962; and Clinch *et al.*, 1972). With varying degrees
of reliability, toxicity to various kinds of animal (including
human beings) is attributed to nectar (or honey derived from nectar) produced by *Aesculus californica, Angelica triquetra,
Asclepias* spp., *Astragalus lentuginosus, Corynocarpus laevigata,
Euphorbia* spp., *Gelsemium sempervirens, Kalmia polifolia, Ledum
palustre, Paullinia* spp., *Rhododendron ponticum, Sophora microphylla, Spirostachys johnstonii* and *Veratrum californicum*. In the
case of *Angelica triquetra*, Bell (1971) describes the effect on
bumblebees (*Bombus* spp.) and other Hymenoptera as narcotic.

     In some cases, a toxic substance has been isolated, such as
the glycoside arbutin (glucose + hydroquinone) from *Arbutus unedo*
honey (Pryce-Jones, 1944). Kozlova (1957) identified another glycoside in the nectar of *Ledum palustra*, and Carey *et al.* (1959)
obtained acetylandromedol from the nectars of species of *Rhododendron*. Very recently, Clinch *et al.* (1972), in New Zealand, have

produced circumstantial evidence that an alkaloid in the nectar of some trees of *Sophora microphylla* is highly toxic to honeybees. As many of the plant families containing the species referred to above are known to be rich sources of alkaloids, it seemed desirable to screen nectars available to us for the presence or absence of these chemicals.

There is some problem in connection with the definition of an alkaloid beyond the general agreement that it is a plant product with a basic character, containing heterocyclic nitrogen, usually synthesized from amino acids or their immediate derivatives (see Hegnauer, 1963). Hegnauer (1963, p. 392) adds that alkaloids are "more or less toxic substances which act primarily on the central nervous system." However, by some definitions, even substances such as thiamin (Vitamin B_1) can be considered alkaloidal (and Ziegler, Lüttge and Lüttge, 1964, found this "favorable" substance to be present in a number of different nectars).

In the present study, two tests for "alkaloids" were employed. To begin with, nectars from 126 species were given the iodoplatinate test described by Smith (1969, p. 519) in which a blue color signifies the presence of compounds containing a tertiary amine group while a white color indicates secondary amine groups and some with primary amine groups. The 19 species giving positive scores here were then retested with Dragendorff's reagent (bismuth nitrate in acetic acid, with aqueous potassium iodide) where a change from pale yellow to orange coloration is a positive result (see Stahl, 1969, p. 873). The eight species which gave positive results on both tests are: *Campanula rapunculoides Cucurbita pepo*, *Cuscuta salina*, *Iris pseudocorus*, *Lotus corniculatus*, *Mimulus moschatus*, *Nymphoides peltatum* and *Rhododendron ponticum*.

Table 9 shows that the sample of plants utilized was a fair one as far as representation of the various pollinator types is concerned; on the other hand, the mean "histidine scale" score of the species giving positive "alkaloid" reactions is considerably higher than that of the "negatives" ($\bar{x}$ = 6.6 versus $\bar{x}$ = 5.0). In addition to having high amino acid concentrations, it is remarkable that six out of the eight species whose nectars give positive "alkaloid" scores also give positive results for lipids *and* for antioxidants. The exceptions are *Lotus corniculatus* and *Rhododendron ponticum*. Five of the eight nectars were also tested for proteins with three positive results. So, once again, there is a notable tendency for nectar-constituents to be positively associated.

The "alkaloid" results are particularly interesting in terms of pollinators involved. All of the plants giving positive results are usually pollinated by bees. On the other hand, not one

of the 46 nectars from species categorized as pollinated by Lepidoptera gave a positive reaction, suggesting that butterflies and
moths do not tolerate the presence of alkaloids in the nectar
they drink. In this respect, the negative result with the nectar
of *Nicotiana sylvestris* is particularly interesting because sap
from the flower-stalk and even the exudate from the stigmatic surface (neither of which is usually absorbed by the moths that
drink the nectar from the flowers) both give a strong positive
reaction.

The apparent sensitivity of anthophilous adult Lepidoptera
to alkaloids (which should be tested by feeding experiments) is
also of interest because the larvae of several groups of butterflies, at least, are able to eat alkaloid-containing tissues with
safety. Thus, there is a close association between the larvae of
the Ithomiinae and the alkaloid-rich Solanaceae, and of larvae of
the Danainae with the alkaloid-rich Apocynaceae and Aristolochiaceae, while the larvae of the Triodini also eat leaves of the
Aristolochiaceae (Ehrlich and Raven, 1964). In this connection,
Gordon (1961, p. 38) makes the interesting suggestion that
"Natural selection...seems to have endowed the larval forms of
many insects, which must undergo variable and severe biochemical
stress, with less vulnerable nervous systems and more active defense mechanisms than the adults, and this 'endurance vigor' is
of value in survival."

Perhaps a little rank speculation is not entirely out of
order near the end of a paper such as this. If so, we might
point out that, many years ago, Robertson (1926) suggested that
butterflies were not responsible for the selection of the basic
characteristics of "butterfly" flowers but have appropriated
flowers that evolved under the influence of long-tongued bees.
If bees do have more resistance to nectar "alkaloids," as is suggested by our results, it might be that there has been selection
in some flowers for the production of protective chemicals against
the flower-inconstant lepidopterans and in favor of the more nearly
oligotropic bees?

Among "secondary" plant products that may help to protect
their producers from attack by phytophagous animals, the so-called
"non-protein" amino acids have received considerable attention
recently (e.g. Rehr, Janzen and Feeny, 1973). γ-aminobutyric
acid is known to affect the functioning of nerve synapses in a
range of invertebrates and vertebrates (see review in Roberts *et
al.*, 1960). It is of very frequent occurrence in vegetative
plant tissue (Fowden, 1962) and, in *Petunia axillaris*, it has
been found in pollen (Linskens and Schrauwen, 1969) and stigma-
exudate (Konar and Linskens, 1966). We have tentatively identified it in the nectar of *Liriodendron tulipiferum* (which nectar
contains a rather unusual assemblage of amino acids that will be

reported upon elsewhere).

Clearly, much more work is needed on this subject of substances potentially *unfavorable* to flower-visitors, and this will require collaboration between researchers qualified in, at least, the fields of organic chemistry, insect toxicology, insect nutrition and botany.

CONCLUSION

This is a synoptic study and, consequently, it has uncovered many problems for which there has not been time to do more than suggest answers. However, we believe that the questions raised in the Introduction relating to the significance for flower-visitors of nectar-constituents have received at least partial illumination.

There is a tendency for nectars that are strong in sugars also to be strong in amino acids and for those that are strong in amino acids to be more frequently the ones that contain detectable lipids and/or antioxidants. When alkaloids occur, it appears that the nectar is usually rich in amino acids and likely to contain lipids, antioxidants and even proteins. This tendency for the concentrations of chemically unrelated substances to be positively associated may stem from what has recently been learned about the mechanism of nectar secretion through the electron microscope studies by Rachmelivitz and Fahn (1973). These authors have shown that the exudation of nectar includes the active extrusion through cellular membranes of vesicles. Such vesicles may contain several chemically unrelated substances. Nevertheless we can say that not all possible substances are included in the vesicles so, to some extent, the nectaries do act as "filters." Some sugars find their way into almost every nectar, and amino acids of one kind or another are almost always included, but lipids and antioxidants are much less frequently included, while proteins and alkaloidal substances are rather rarely a part of the nectar.

For the consumers of the nectar, it appears that some of the contents are of nutritional value. This is true, without question, for the sugars. The amino acids are concentrated enough and appropriate enough in their range of identities to be valuable in those cases where there is no abundant supply from other sources (most Lepidoptera) or where insects must be lured away from some other source (as with carrion- and dung-flies). Nocturnally produced nectars appear to be inherently rather weaker in amino acids than those produced diurnally, but this may be compensated for by the uptake of a relatively large volume of nectar (notably by hovering moths). The lipids may be nutritionally useful to bees and flies that can digest them and, additionally, may be of significance in providing flight-energy for very actively flying hummingbirds and hawkmoths. Here, much new work is needed. On the other hand, proteins, where they have been detected, seem more likely to have an enzymic significance than to serve as food for flower-visitors. The antioxidants, when they are present, may also play "chemical" rather than "nutritional" roles in nectar, preventing the destruction by oxidation of other substances. Some other chemicals, such as alkaloids, non-protein amino acids and glycosides, may be protective in a different sense, favoring continued flower-visits by more efficient pollinators through deterrence of less favorable ones. The extent of the interplay between attractants and deterrents in determining which animals visit flowers, and the relationships of these visits to the breeding systems of the plants are in need of further investigation.

To follow the synoptic study reported on here, two kinds of more detailed research are proposed. The first involves investigation by experimental means of individual pollinator-flower relationships, especially in relation to the amounts of chemicals that the visitors ingest and their need for or sensitivity to those chemicals. The second is to place the relationships in their appropriate ecosystem contexts. Thus, the failure of any one flower species to provide a complete range of essential and quasi-essential amino acids in nectar means that if an adequate and balanced supply is to be obtained by a nectar-imbiber to whom it may be valuable, several flower species must be visited. We are now investigating the availability of such nutritional necessities during the seasons of activity and within the flying ranges of the animals concerned in several ecosystems in California and Costa Rica.

APPENDIX I

Species whose nectar gave positive indications of proteins: *Alstroemeria haemantha, Bergenia latifolia, Beschorneria yuccoides, Brodiaea volubilis, Calluna vulgaris, Colletia cruciata, Convolvulus arvensis, Cucurbita pepo, Dicentra formosa, Erica mediterranea, Erysimum concinnum, Fremontodendron californicum, Iris pseudacorus, Lilium humboldtii, Menyanthes trifoliata, Piaranthus pillansii* var. *inconstans, Polygonum cocineum, Silene dioica, Zigadenus fremontii.*

APPENDIX II

Species whose nectar gave positive indications of lipids: *Abronia latifolia, Aesculus californica, Aquilegia eximia, A.* sp.

(hort.), *Arctostaphylos uva-ursi, Brodiaea laxa, Buddleia globosa, Callistemon* sp., *Calochortus uniflorus, Campanula rapunculoides, Castilleja affinis, Catalpa bignonioides, Clarkia elegans, Convolvulus arvensis, C. soldanella, Coreopsis gigantea, Corydalis lutea, Cotoneaster franchettii, Crataegus phaenopyrum, Crinum moorei, Cucurbita pepo, Cuscuta salina, Digitalis viridiflora, Echeveria* sp., *Echinopsis turbinata, Eriodictyon californicum, Ervatamia coronaria, Erysimum concinnum, Frankenia grandifolia, Fritillaria lanceolata, Fuchsia arborescens, Geranium robertianum, Gladiolus* sp., *Grevillea obtusifolia, Heuchera micrantha, Impatiens sultanii, Iris douglasiana, I. pseudacorus, I.* sp. (hort.), *Jacaranda acutifolia, Jaumea carnosa, Lilium maritimum, Limnanthes douglasii* var. *sulphurea, Lythrum californicum, Mamillaria theresae, Matthiola incana, Menyanthes trifoliata, Mertensia echioides, Mesembryanthemum edule, Mimulus aurantiacus, M. cardinalis, M. moschatus, Nigella damascena, Nothoscordum fragrans, Nymphoides peltatum, Piaranthus pillansii* var. *inconstans, Pittosporum rhombidifolium, Potentilla egedii* var. *grandis, P. glandulosa, Proboscidea louisianica, Prunus lusitanica, Ranunculus repens, Rhododendron* sp. (hort.), *Rhus typhina, Ribes speciosum, Rubus vitifolius, Salvia mellifera, Sedum dendroideum, Silene dioica, Sorbus* sp., *Thymus serpyllum, Tiarella unifoliata, Trichocereus andalgalensis, Trifolium wormskioldii, Vicia faba.*

APPENDIX III

Species whose nectar gave positive indications of antioxidants: *Achillea millefolium, Aesculus californica, Agave toumeyana, Allium peninsulare, Alstroemeria haemantha, Aquilegia coerulea, A. eximia, Arenaria douglasii, Armeria maritima, Buddleia globosa, Calochortus uniflorus, Campanula rapunculoides, Catalpa bignonioides, Clarkia elegans, Clivia miniata, Cucurbita pepo, Cuscuta salina, Dianthus barbatus, D. plumarius, Digitalis purpurea, Drimys winteri, Dudleya edulis, Echeveria* sp., *Epilobium angustifolium, Eriogonum* sp., *Eriophyllum confertiflorum, Erysimum concinnum, Frankenia grandifolia, Geranium robertianum, Iris pseudacorus Iris* sp. (hort.), *Kentranthus ruber, Lavandula spicata, Lilium humboldtii, L. maritimum, Limnanthes douglasii* var. *sulphurea, Ludwigia palustris* var. *pacifica, Lychnis chalcedonica, Mamillaria theresae, Medicago sativa, Menyanthes trifoliata, Mimulus cardinalis, M. moschatus, Navarretia squarrosa, Nicotiana tabacum, Nymphoides peltatum, Oenothera organensis, Oxalis rubra, Penstemon cordifolius, Piaranthus pillansii* var. *inconstans, Potentilla glandulosa, Proboscidea louisianica, Ranunculus repens, Ribes speciosum, Rubue vitifilius, Salvia mellifera, Sedum dendroideum, Sidalcea malvaeflora, Silene dioica, Thymus serpyllum,*

Trichocereus andalgalensis, *Trifolium wormskioldii*, *Vicia faba*.

LITERATURE CITED

Agthe, C. 1951. Über die physiologische Herkunft des Pflanzen-nektars. Ber. schweiz. bot. Ges. 61:240-274.

Auclair, J. L. and Jamieson, C. A. 1948. A qualitative analysis of amino acids in pollen collected by bees. Science 108: 357-358.

Bailey, L. 1954. The filtration of particles by the proventriculi of various Aculeate Hymenoptera. Proc. roy. ent. Soc. Lond. 29:119-123.

Baker, H. G. 1973. Evolutionary relationships between flowering plants and animals in American and African forests. *In* Tropical forest ecosystems in Africa and South America: a comparative review (B. J. Meggers, E. S. Ayensu and W. D. Duckworth, eds.), pp. 145-159 (Chap. 11), Washington, D. C.

Baker, H. G. and Baker, I. 1973a. Amino acids in nectar and their evolutionary significance. Nature 241:543-545.

Baker, H. G. and Baker, I. 1973b. Some anthecological aspects of the evolution of nectar-producing flowers, particularly amino acid production in nectar.   Taxonomy and ecology (V. H. Heywood, ed.), pp. 243-264 (Chap. 12), London and New York.

Baker, H. G., Baker, I. and Opler, P. A. 1974. Stigmatic exudates and pollination. Pollination and dispersal (N. B. M. Brantjes, ed.), pp. 47-60, Univ. of Nijmegan.

Bänziger, H. 1971. Bloodsucking moths of Malaya. Fauna 1:4-16.

Becker, J. and Kardos, R. F. 1939. Ueber den Vitamin-C-Gehalt von Honig. Z. Untersuch. Lebensmitt. 78:305-308.

Bell, C. R. 1971. Breeding systems and floral biology in the Umbelliferae. *In* The biology and chemistry of the Umbelliferae (V. H. Heywood, ed.), pp. 93-107, London.

Beutler, R. 1930. Biologisch-chemische Untersuchungen am Nektar von Immenblumen. Z. vergl. Physiol. 12:72-176.

Beutler, R. 1953. Nectar. Bee World 24:106-116;128-136;156-162.

Bonnier, G. 1878. Les nectaires. Ann. Sci. natur., VI ser.,
    Bot., 8:5-212.

Bukatsch, F. and Wildner, G. 1956. Ascorbinsäurebestimmung in
    Nektar, Pollen, Blütenteilan und Elaiosomen mit Hilfe einer
    neuen Mikromethode. φYTON 7:34-46.

Butler, C. G. 1954. The world of the honeybee. London.

Buxbaum, F. 1927. Zur Frage des Eiweissgehaltes des Nektars.
    Planta (Berl.) 4:818-821.

Cain, A. J. 1950. The histochemistry of lipoids in animals.
    Biol. Revs. 25:73-112.

Clinch, P. G., Palmer-Jones, T. and Forster, I. W. 1972. Effect
    on honeybees of nectar from the Yellow Kowhai (Sophora
    microphylla Ait.). N.Z. J. Agric. Res. 15:194-201.

Cotti, T. 1962. Über die quantitative Messung der Phosphataseak-
    tivität in Nektarien. Ber. schweiz bot. Ges. 72:306-331.

Dadd, R. H. 1972. Insect nutrition: current developments and
    metabolic implications. Ann. Rev. Entom. 18:381-420.

Echigo, T. 1971. Studies on relationship of chemical components
    of honey, nectar and pollen. Bull. Fac. Agric. Tamagawa
    Univ. 11:37-54.

Ehrlich, P. R. and Raven, P. H. 1964. Butterflies and plants:
    a study in coevolution. Evolution 18:586-608.

Ewert, R. 1932. Die Nektarien in ihrer Bedeutung für Bienenzucht
    und Landwirtschaft. Leipzig.

Faegri, K. and van der Pijl, L. 1971. The principles of pollina-
    tion ecology (2nd edition). Oxford.

Fahn, A. 1949. Studies in the ecology of nectar secretion.
    Palest. J. Bot., Jerusalem 4:207-224.

Ford, E. B. 1945. Butterflies. London.

Fowden, L. 1962. The non-protein amino acids of plants.
    Endeavour 21:35-45.

Free, J. B. 1970. Insect pollination of crops. London and New

York.

Free, J. B. and Butler, C. G.   1959.   Bumblebees. London.

Gilbert, L. E.   1972.   Pollen feeding and reproductive biology of
    Heliconius butterflies.   Proc. Nat. Acad. Sci. U.S.A. 69:
    1403-1407.

Gilmour, D.   1961.   The biochemistry of insects.   New York.

Gilmour, D.   1965.   The metabolism of insects.   Edinburgh and
    London.

Gontarski, H.   1948.   Ein Vitamin C oxydierendes Ferment der
    Honigbiene.   Z. Naturf. 3b(7/8):245-249.

Gordon, H. T.   1961.   Nutritional factors in insect resistance to
    chemicals.   Ann. Rev. Entom. 6:27-54.

Grant, K. A. and Grant, V.   1968.   Hummingbirds and their flowers.
    New York and London.

Griebel, C. and Hess, G.   1939.   Vitamin C-enthaltende Honige.   Z.
    Untersuch. Lebensmitt. 78:308-314.

Griebel, C. and Hess, G.   1940.   The vitamin C content of flower
    nectar of certain Labiatae (in German).   Z. Untersuch. Leben-
    smitt. 79:168-171 (Chem. Abstr. 34: No. 44187, 1940).

Haydak, M. H.   1970.   Honey bee nutrition.   Ann. Rev. Entom. 15:
    143-156.

Haydak, M. H., Palmer, L. S., Tanquary, M. C. and Vivano, A. E.
    1942.   Vitamin contents of honeys.   J. Nutrit. 23:581.

Hegnauer, R.   1963.   Comparative phytochemistry of alkaloids.
    In Comparative phytochemistry (T. Swain, ed.), pp. 211-230
    (Chap. 13), London and New York.

Heinrich, B.   1971.   Thoracic temperature regulation in the
    sphinx moth, *Manduca sexta*.   Parts I and II.   J. Exp. Biol.
    54:141-166.

Hocking, B.   1953.   The intrinsic range and speed of flight of
    insects.   Trans roy. ent. Soc. Lond. 104:223-345.

Hocking, B.   1968.   Insect-flower associations in the high Arctic with special reference to nectar. Oikos 19:359-387.

Hocking, B. 1971. Six-legged science. Cambridge, Mass.

House, H. L. 1965. Insect nutrition. *In* The physiology of the
Insecta, Vol. 2 (M. Rockstein, ed.), pp. 769-857, New York.

Hutchinson, J. 1959. The families of flowering plants (2nd edition). Oxford.

Jaeger, P. 1961. The wonderful life of flowers. New York.

Kartashova, N. N. and Novikova, T. N. 1964. A chromatographic
study of the chemical composition of nectar (in Russian).
Isv. Tomskogo Otd. Vses Botan. Obshchestva 5:111-119. (Chem.
Abstr. 64: No. 8641a, 1966).

Kask, M. 1938. Vitamin C - Gehalt der estnischen Honigen. Z.
Untersuch. Lebensmitt. 76:543-545.

Kirchner, O. 1911. Blumen und Insekten. Leipzig.

Klots, A. B. 1958. The world of butterflies and moths. New
York.

Knuth, P. 1906-1909. Handbook of flower pollination. (Translated by J. R. A. Davis). Oxford.

Konar, R. N. and Linskens, H. F. 1966. Physiology and biochemistry of the stigmatic fluid of *Petunia hybrida*. Planta
(Berl.) 71:372-387.

Kozhantshikov, I. W. 1938. Carbohydrate and fat metabolism in
adult Lepidoptera. Bull. Ent. Res. 29:103-114.

Kozlova, M. V. 1957. Nectar of *Ledum palustre* as a possible
source of the toxicity of honey (in Polish). Voprosy
Pitaniya 16:80. (Chem. Abstr. 52: No. 5695g, 1958).

Leppik, E. E. 1953. The ability of insects to distinguish number. Amer. Natur. 87:228-236.

Linskens, H. F. and Schrauwen, J. 1969. The release of free
amino-acids from germinating pollen. Acta bot. neerl. 18:
605-614.

Lodder, J. and Krieger-van Rij, N. J. W. 1952. The yeasts: a taxonomic study. New York.

Lüttge, Ü. 1961. Über die Zusammensetzung des Nektars und den Mechanismus seiner Sekretion. I. Planta (Berl.) 56:189-212.

Lüttge, Ü. 1962a. Über die Zusammensetzung des Nektars und den Mechanismus seiner Sekretion. II. Planta (Berl.) 59:108-114.

Lüttge, Ü. 1962b. Über die Zusammensetzung des Nektars und den Mechanismus seiner Sekretion. III. Planta (Berl.) 59: 175-194.

Lüttge, Ü. 1966. Funktion und Struktur pflanzlicher Drüsen. Naturwissenschaften 53:96-103.

Maslowski, P. and Mostowska, I. 1963. Electrochromatographic estimation of free amino acids in honeys (in Polish). Pszczel. Zeszyty Nauk 7:1-6. (Chem. Abstr. 60: No. 2258c, 1964).

Mauritzio, A. 1945. Giftige Bienenpflanzen. Beih. schweiz. Bienenztg. 47:178-182.

Meeuse, B. J. D. 1961. The story of pollination. New York.

Mostowska, I. 1964. Amino acids of nectars and honeys (in Polish). Zeszyty Nauk Wyzszej. Szkoly Rolniczej Olsztynie 20:417-432. (Chem. Abstr. 64: No. 20529, 1966).

Müller, H. 1883. The fertilisation of flowers (translated and edited by D. W. Thompson). London.

Nair, A. G. R., Nagarajan, S. and Subramanian, S. 1964. Chemical compositions of nectar in Thunbergia grandiflora. Current Sci. (India) 33:401.

Niethammer, A. 1930. Mikrochemie einzelner Blüten in Zusammenhang mit der Honiggewinnung. Gartenbauwiss. 4:85-98.

Palmer-Jones, T. and Line, L. J. S. 1962. Poisoning of honey bees by nectar from the karaka tree. N.Z. J. Agric. Res. 5:433-436.

Percival, M. S. 1961. Types of nectar in angiosperms. New Phytol. 60:235-281.

von Planta, A.   1886.   Ueber die Zusammensetzung einiger Nekterar-
    ten.   Hoppe-Seyl. Z. 10:227-247.

Proctor, M. and Yeo, P.   1973.   The pollination of flowers.
    London.

Pryce-Jones, J.   1944.   Some problems associated with nectar,
    pollen and honey.   Proc. Linn. Soc. Lond. 1944:129-174.

Pryce-Jones, J.   1950.   The composition and properties of honey.
    Bee World 31:2-6.

Rachmelivitz, T. and Fahn, A.   1973.   Ultrastructure of nectaries
    of *Vinca rosea* L., *Vinca major* L. and *Citrus sinensis* Osbeck
    cv. *Valencia* and its relation to the mechanism of nectar
    secretion.   Ann. Bot. (Lond.) 37:1-9.

Rehr, S. S., Janzen, D. H. and Feeny, P. P.   1973.   L-dopa in
    legume seeds: a chemical barrier to insect attack.   Science
    181:81-82.

Roberts, E., Baxter, C. F., Harreveld, A. V., Wiersma, C. A. G.,
    Adey, W. R. and Killam, K. F. (eds.)   1960.   Inhibition in
    the nervous system and gamma-aminobutyric acid.   New York.

Robertson, C.   1926.   Quoted from M. S. Percival (1961) - *vide
    supra*.

Robertson, C.   1928.   Flowers and insects.   Carlinville, Ill.

Rychlik, M. and Federowska, Z.   1963.   Specific optical rotation
    of nectar dextrins (in Polish).   Pszczel. Zoszyty Nauk 7:
    7-14 (Chem. Abstr. 59: No. 13113g, 1963).

Sandholm, H. A. and Price, R. D.   1962.   Field observations on
    the nectar feeding habits of some Minnesota mosquitoes.
    Mosquito News 22:346-349.

Shuel, R. W.   1955.   Nectar secretion.   Amer. Bee J. 95:229-234.

Smith, I.   1969.   Chromatographic and electrophoretic techniques,
    vol. I.   Chromatography (2nd edition).   London.

Sporne, K. R.   1969.   The ovule as an indicator of evolutionary
    status in angiosperms.   New Phytol. 68:555-566.

Sprengel, C. K.   1793.   Das Endekte Geheimnisse der Natur im Bau und in der Befruchtung der Blumen. Berlin.

Stahl, E. 1969. Thin-layer chromatography (2nd edition) (translated by M. R. F. Ashworth). Berlin.

Thien, L. B. 1969. Mosquito pollination of *Habenaria obtusa* (Orchidaceae). Amer. J. Bot. 56:232-237.

Ulrich, W. 1938. Die Entdeckung eines vitaminhaltigen Honigs (Vitamin C). Ber. 7 Int. Congr. Ent.: 1883-1887.

Vogel, S. 1969. Flowers offering fatty oil instead of nectar. Abstracts XI Int. Bot. Congr., Seattle, p. 229.

Waller, G. D. 1973. Chemical differences between nectar of onion and competing plant species and probable effects upon attractiveness to pollinators. Ph.D. Thesis, Utah State University.

Weber, F. 1942. Vitamin C in Nektar von Fritillaria imperialis. Protoplasma 36:316.

Weber, F. 1951. Impatiens-Nektar. Phyton (Horn) 3:110-111.

Wigglesworth, V. B. 1966. Insect physiology (6th edition). London.

Young, A. M. 1972. Community ecology of some tropical rain forest butterflies. Amer. Midl. Nat. 87:146-157.

Zalewski, W. 1966. Phosphatases in honey (in Polish). Pszczel. Zeszygty Nauk 9:1-34 (Chem. Abstr. 64: No. 10325g, 1956).

Zauralov, O. A. 1969. Oxidizing enzymes in nectaries and nectar (in Russian). Tr. Nauch.-Issled Inst. Pchelovod. 1969:197-225 (Chem. Abstr. 74: No. 11050, 1971).

Zebe, E. C. 1953. Über den respiratorischen Quotienten der Lepidopteren. Naturwissenschaften 40:298.

Zebe, E. C. 1954. Über den Stoffwechsel der Lepidopteren. Z. vergl. Physiol. 36:290-317.

Zebe, E. C. 1959. Discussion. Fourth Int. Congr. Biochem. 12: 198-199.

Ziegler, H. 1956. Untersuchungen über die Leitung und Sekretion der Assimilate.   Planta (Berl.) 47:447-500.

Ziegler, H. and Lüttge, U.   1959.   Über die Resorption von C^{14}
Glutaminsaure durch sezernierende Nektarien.   Naturwissenschaften 46:176-177.

Ziegler, H., Lüttge, U. and Lüttge, U.   1964.   Die wasserlösichen
Vitamine des Nektars.   Flora (Jena) 154:215-229.

Zimmerman, J. G.   1953.   Papierchromatographische Untersuchungen
über die pflanzliche Zuckersekretion.   Ber. schweiz bot.
Ges. 63:402-429.

Zimmerman, J. G.   1954.   Über die Sekretion saccharosespaltender
Trans-glucosidasen in pflanzlichen Nektar.   Experientia
(Basel) 9(3):145-149.

THE ROLE OF ENERGETICS IN BUMBLEBEE-FLOWER INTERRELATIONSHIPS

Bernd Heinrich

Division of Entomology
University of California
Berkeley, California 94720

INTRODUCTION

One of the potentially useful ways of examining the interactions of organisms with their environment is from the standpoint of energetics. Such considerations are particularly rich in
implications with respect to pollinators, some of which derive all
of their food energy from the sugar in the nectar of flowers. The
energetics of the pollinators can be viewed with regard to foraging strategies, the fertilization of flowers, gene flow, species
interaction, as well as from the standpoint of coevolution.

Foraging behavior of pollinators (Macior, 1971) and possible
correlations of the energetics of the pollinators to numerous
aspects of pollination biology have recently been discussed
(Heinrich and Raven, 1972). It is obvious that the role of
energetics must be examined in the context of the natural environment with its native pollinators, or under controlled conditions.
Few extrapolations from one situation to the next will be possible
until more quantitative data on the energy costs and pay-offs of
foraging are available.

I will here examine foraging energetics of bumblebees in one
ecological arena where they are common native pollinators throughout the flowering season. The possible energy pay-offs and the
costs while foraging from some of the common flowers with different
amounts of food reward, and different densities, will be examined.
The results may indicate to what extent energetics dictates the
foraging behavior.

MATERIALS AND METHODS

The observations refer to Franklin County, Maine. The area
contains deciduous forests, farmlands, bogs, and small mountains
with evergreen forest. The movements of the bees was timed with a
stopwatch. Nectar was collected from flowers in 1 μl to 10 μl
capillary tubes, and the volume was read to the nearest mm. The
concentration of the nectar was read with a Bellingham and Stanley
pocket refractometer reading to 50% sugar concentration. Higher
nectar concentrations were determined after dilution. The corolla
of many of the flowers was too narrow to insert the 1 μl capillary

tube.   In these flowers nectar was removed by squeezing the base
of the flower between thumb and forefinger.

RESULTS

Floral sugar: production and availability

A first step in an evaluation of the role of energetics in
the foraging behavior is a quantification of the food rewards of
the flowers visited.   A distinction must here be made between what
is available in the flowers and what they produce.   The common
flowers visited by bumblebees during May-June produce a wide range
of sugar amount per floret (Fig. 1).   In early spring and late
summer most of this sugar is removed by the bees, but there is a
surplus of nectar for some brief periods during mid-summer
(Heinrich, unpublished).

Although the nectar production of the flowers varies greatly,
the amount of sugar available in a variety of unscreened flowers
tends to be similar.   The foragers remove the sugar until, on the
average, one or two tenths of a mg remain by mid-morning (Fig. 1).
The few species containing more than this amount were flowers that
were relatively rare or blooming in small, widely separated patches.

The amounts of food per floret, as such, obviously means little in terms of the energetics of the foragers until flower
distribution and rate of flower visitation (to be discussed) are
known.   In energetic terms it is possible that the food rewards
are more uniform than Figure 1 may indicate, for flowers 16-24
(Fig. 1) occur in inflorescences of hundreds or thousands of
blossoms, in trees or bushes with thousands of flowers, and/or in
dense colonies.   Most of the flowers of greater nectar rewards
(numbers 1-15) occur on widely dispersed plants and/or on plants
with relatively few flowers per plant, or with fewer flowers per
inflorescence.

In terms of energetics the average amount of food per flower,
and the rate at which they can be visited, are interrelated.   The
absolute amount of food in any one floret, as such, may be of
little significance since the bees are usually rewarded with food
at only a small percentage of flowers that they visit.   This is
because (1) the variability of nectar production of individual
florets may be great (Fig. 2), and (2) as the bees forage throughout the day, the number of unvisited flowers decreases while the
buildup of nectar in these flowers continues to increase.

Speed of foraging

The distribution of flowers on the plant, and their distribution in the environment, will dictate, in large part, how many can

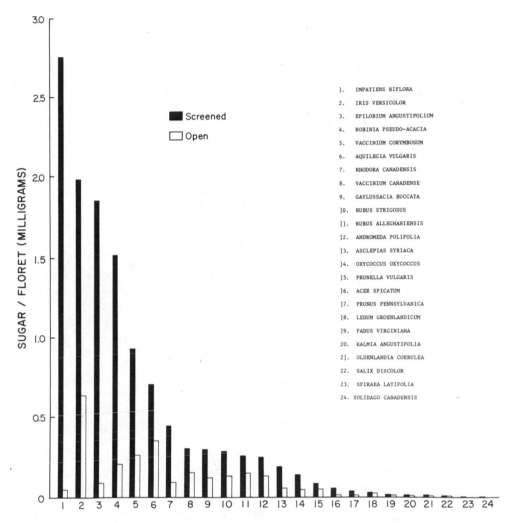

Figure 1. The mean amounts of sugar in screened (for 24 hours) and unscreened flowers utilized by bumblebees. In most cases the sample sizes for each bar are 20-30. The nectars were sampled in the forenoon.

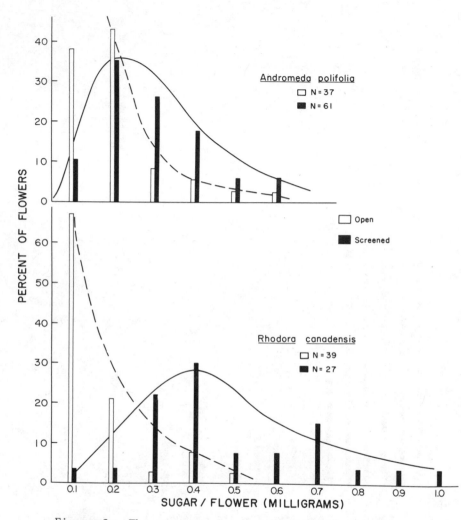

Figure 2. The amounts of sugar in individual blossoms of two species of flowers blooming side by side in a bog. *Andromeda polifolia* was less abundant than *Rhodora canadensis*. Filled bars = flowers screened for 24 hours. Open bars = unscreened flowers. Lines were gauged by eye.

be visited per unit time. This may shift the balance between profit and loss to an animal foraging at flowers with given food
rewards.

Most of the flowers from which bumblebees were foraging in
central Maine occur in colonies. The flowers may be in "real"
colonies in that the flowers are on different plants, or the
flowers may be clumped by being on the same tree or bush. In both
cases the factors affecting profits to the bee remain the same,
although the factors affecting cross-pollination are different.

The bees are conditioned to forage at specific flowers in
specific sites (Manning, 1956). It is unlikely that they would
continue to forage at a site where they do not make a profit.
This would set limits to the minimum permissable flower density at
their foraging sites. Perhaps in part or as a result of this
restriction, the number of flowers of a given species visited per
unit time is similar from one site to the next. In order to
further investigate the energetics of foraging the speed at which
various classes of flowers were visited has been recorded (Table
1).

Of the various taxa of bees in the area, the social bumblebees visit the most flowers per unit time. Relatively single
blossoms tend to be visited at approximately 20 per minute (Table
1). Relatively clumped blossoms, such as *Lonicera*, are visited at
the rate of 40 per minute while the florets on a clover inflorescence may be probed at approximately 50 per minute. Blossoms in
foliage, such as those of *Rubus strigosus* and *Impatiens biflora*,
require more search and are visited at lower rates. The bees
spend approximately a second at most open blossoms, but inflorescences of *Hieracium*, for example, retain the queens of *B. vagans*
for an average of 2.1 seconds. The solitary bees of numerous
species visit about one-third as many flowers per minute as
Bombus at the higher ambient temperatures. The non-social bumblebees (*Psithyrus*) are conspicuously slower than *Bombus* (Heinrich,
unpublished).

Ease of entrance into the blossoms, as well as flower distribution in the colony, greatly affect foraging rate. The relatively small long-tongued (11.7 mm) *B. vagans* queens, for example,
enter the bell-shaped flowers of *Uvularia sessifolia* rapidly,
spending 4.8 seconds at each blossom and visiting 11 per minute at
the plant colonies. The short-tongued (9.4 mm) *B. terricola*
infrequently visit these flowers. (One that did so spent an
average of 10.3 seconds per flower). *Bombus ternarius* (mean
tongue length = 9.9 mm), which did not visit the flowers of *U.
sessifolia* as frequently as *B. vagans*, was slower at these flowers,
visiting only 6.5 per minute. The high metabolic rate of the
bumblebees, and the inability of the short-tongued bees to gain
rapid access to the nectar, should be a large factor determining

Table 1. The rates at which queens of *Bombus vagans* visited various types of flowers.

	Flower arrangement	"Flowers"/Min.			Plants/Min.		
		$\overline{X}$	N	S.E.	$\overline{X}$	N	S.E.
Rhodora canadensis (at 11°C)	Relatively single, exposed	19.7	(16)	1.5	–	–	–
Rhodora canadensis (at 22°C)	Relatively single, exposed	19.6	(20)	1.0	4.2	(14)	0.6
Vaccinium corymbosum	Relatively singl , exposed	19.0	(12)	1.4	2.7	(17)	0.3
Rubus strigosus	Single, in foliage	11.4	(20)	0.7	–	–	–
Uvularia sessifolia	Single, in foliage	11.0	(10)	1.0	11.0	(10)	1.0
Hieracium sp.	Inflorescence	12.8	(5)	1.7	7.6	(5)	0.9
Iris versicolor	Widely spaced, exposed	5.4	(8)	1.0	3.6	(8)	0.8
Salix sp.	Inflorescence	2.8	(13)	0.7	–	–	–
Lonicera sp.	Tightly clumped, exposed	38.8	(22)	1.0	< 0.2	(6)	0.1

their foraging profits.

The speed of foraging of the bees depends markedly on their body temperature. When the supply of nectar is ample, bumblebees usually maintain a high body temperature and visit nearly the same number of flowers per minute regardless of air temperature (Table 1). The solitary bees, however, seldom forage at air temperatures below 15 C, and when they do so they visit relatively few blossoms per unit time, spending most of their time in torpor.

Energy expenditure and caloric profits of foraging

Given the energetic costs of foraging, the significance of a certain food reward of a flower can be examined in relation to its pollinators. Computations of foraging costs and profits can be used to indicate which flowers in the environment may be suitable sources of nectar, and under what conditions of temperature and flower distribution they become unsuitable. It may be possible to make predictions of the foraging behavior, and to determine the strategies available to the plant to maximize out-crossing while minimizing nectar production.

Stationary bumblebees sometimes have an energy expenditure at low air temperatures as high as that in flight. The elevated metabolic rate helps to maintain a high body temperature, allowing them to continue foraging (Heinrich, 1972a). On the basis of these energy expenditures, the food calories provided by the flowers (Fig. 1), and the rates of flower visitation (Table 2), it is a simple matter to compute approximately how many flowers must be visited per unit time to defray the energetic cost of foraging (Fig. 3).

Although such calculations must necessarily involve simplifications and assumptions, they strongly suggest that the energetic cost of foraging of the large bees (queens) would exceed the caloric returns that could be gathered from some flowers. For example, in view of observed foraging rates (Table 2) the sugar content of the flowers of *Prunus pennsylvanica* is not sufficient for the bees to make a profit if they forage at low air temperature (Fig. 3). It is possible, however, that workers, weighing approximately one third as much as the queens and visiting approximately the same number of flowers per unit time, would go into energetic debt one third as fast, or make three times more profit, depending on the nectar content of the flower. The calculations also indicate that the effect of temperature is negligable in terms of energy balance while foraging from some flowers, while it would be a major factor determining whether or not a profit can be made on others.

The foragers must do more than merely meet the energetic costs of foraging. At least in the social insects, the animals must make a profit above cost in a certain time limit. The caloric

Table 2. Calculated caloric profits and foraging times of a bumblebee (0.2 g) in terms of the nectars (mg sugar, volume, %
sugar) in flowers (unscreened) of different species. The
calculations (B,C) are based on the observed rates of
flower visitation, on the assumption that the bee spends
50% of its time in flight and fills its honeystomach
(100 µl), and has an energy expenditure of 0.5 cal/min
for flight and for thermoregulation at 0 C (Heinrich,
1972a, and unpublished). The flowers were examined for
nectar in the forenoon, and sample sizes in most cases
ranged between 20 and 30. The first six flowers were
visited primarily by bumblebees. The second six were
visited by numerous solitary bees, and occassionally by
bumblebees at the higher ambient temperatures.

	A Nectar and sugar/flower mg (µl, %)	B Cal profit/min 0 C (24 C)	C Minutes of foraging time 0 C (24 C)
Vaccinium corymbosum	0.274 (1.52, 18)	19.3 (19.8)	3.45 (3.38)
Rhodora canadensis	0.096 (0.30, 32)	6.10 (6.60)	19.4 (17.9)
Iris versicolor	0.625 (2.50, 25)	8.24 (8.74)	11.2 (10.6)
Andromeda polifolia	0.147 (1.05, 14)	9.88 (10.33)	5.25 (5.00)
Oxycoccus oxycoccus	0.048 (0.20, 24)	2.56 (3.06)	34.7 (29.0)
Prunella vulgaris	0.034 (0.17, 20)	2.15 (2.65)	34.5 (28.0)
Rubus strigosus	0.027 (0.15, 18)	0.10 (0.60)	666 (111)
Prunus pennsylvanica	0.0135 (0.05, 27)	0.75 (1.25)	134 (80.0)
Ledum groenlandicum	0.030 (0.04, 75)	2.88 (3.33)	96.2 (83.3)
Padus virginiana	0.0098 (0.041, 24)	0.26 (0.70)	340 (125)
Oldenlandia coerulea	0.008 (0.019, 42)	Loss (0.28)	∞ 266
Kalmia angustifolia	0.0074 (0.074, 10)	Loss (0.185)	∞ 200

profit, and the time required to collect it, can also be computed
from the foraging costs, foraging speeds and the sugar contents
of the flowers (Table 2). Such calculation (disregarding flight
time to and from the foraging site) suggest that the bees should
be able to fill their honeystomach with nectar in several minutes
from some flowers, provided these have not been visited previously. However, the amounts of nectar available in the flowers dictate minimum foraging times of over half an hour in some of the
flowers commonly used by the bees. The bees are willing to continue foraging much longer than this on a single foraging trip.
One worker of *B. fervidus* was continuously observed for 122
minutes. During this time it visited 800 flowers of primarily
two species (Heinrich, unpublished).

Food rewards and pollinator vagility

In order for a plant species to evolve to utilize a given
type forager as a pollinator, the animal must move not only from
flower to flower but also from plant to plant. Presumably that
animal with the greatest vagility between plants of a species has
the best potential to shape the flower's evolution with respect
to morphology, identifying features, food rewards and flowering
time. It is of interest that some of the flowers blooming in the
early spring, and primarily visited by bumblebee queens, have
relatively large amounts of nectar. If they were visited only by
the native solitary bees, weighing 4-60 mg, (approximately 6 to
150 times less than the queens) fewer flowers would be visited
per foraging trip. The observations suggest that the bumblebees
may be, bee for bee, better agents of pollen transfer than the
native solitary bees. The rate of flower visitation of the solitary bee is low, and the bees should be satiated after visiting
only several flowers.
Estimates of the number of flowers visited per foraging trip
indicate that bumblebees could be effective agents of pollen
transfer even in those flowers having relatively large amounts of
nectar. The greatest amounts of nectar available in unshielded
native flowers were observed in *Iris versicolor*, containing on
the average 2.5 μl per floret. It would require the nectar contents of approximately 40 flowers to fill one honeystomach. The
amount of nectar is thus still low enough so that many flowers are
presumably visited and pollinated per foraging trip.
The effect of food rewards on the behavior of the bees may
be more critical in plants with many flowers than those with few.
Since bees are well-known to be highly site-specific in their
foraging behavior (see review, Grant, 1950) this suggests that,
while foraging at a tree with thousands of flowers, they would
remain and thus accomplish no cross-pollination regardless of

the numbers of flowers visited. However, five *Bombus terricola* queens visited few of the approximately 15,000 blossoms on a *Prunus pennsylvanica* tree, remaining to forage on the average only 3.2 seconds before flying on. The smaller *B. ternarius* on the other hand, remained to forage on the average 4.9 minutes (N = 10) at the same tree. The small amounts of nectar of the flowers (Fig. 1), may make it uneconomical (Fig. 3) for the bees to remain site specific, causing them to examine other trees.

Deception and possible energy "parasitism"

It is well known that bumblebees (Manning, 1956), like honeybees (von Frisch, 1967) become conditioned to forage at specific flowers which they identify by their scents, shapes and colors. These signals are "meaningful" only in terms of the food rewards. However, after the food rewards are removed the bees may for a short time search for food at the signals with which the food was originally identified. Preliminary observations suggest that this tendency may have been used to advantage by some flower species to minimize their nectar production. Systematic efforts have not been undertaken, but an analysis of the nectar contents of several species of flowers of similar appearance from the same habitat revealed that the first to bloom in most cases contained more nectar than the second. The duration of most of the bloom in several species was less than two weeks. Perhaps the second species to bloom is, in part, "parasitic" on the first. In order to ascertain if this is so it will be necessary to observe if the same individual foragers visit the similar flowers of different species, and to compare the energy budgets of these foragers on these different flower species.

Figure 3. The calculated cost of foraging in terms of the nectar contents of a variety of flowers. The sample includes those occasionally, as well as those commonly, visited by bumblebees. The calculations are based on the nectar contents of unscreened flowers in the forenoon. The forager is assumed to be an 0.6 g bee (queen) spending 50% of her time in flight and having a metabolic rate of 3.0 cal/min in flight and while regulating its body temperature at 0°C (Heinrich, 1972a and unpublished).

    ** = visited at both low and high temperatures by the large bees in the study area
     * = visited at higher temperatures
     · = occasionally visited at higher temperatures

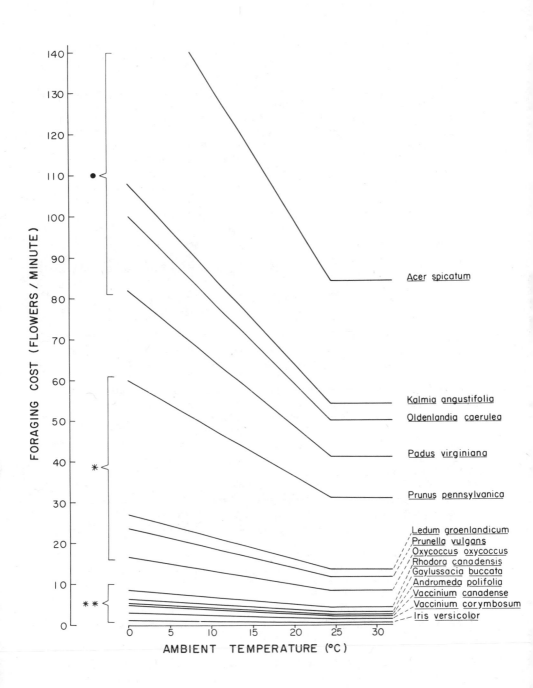

In addition, it is possible that some flowers attract naive pollinators seeking food without providing it. A showy purplish-pink orchid, *Calopogon pulchellum*, blooming in bogs from Newfoundland to Florida, provides neither nectar nor pollen. In comparison to other flowers, there was almost no insect activity at several hundred of these flowers in a clearing on a sphagnum bog. However, during two afternoons of observation a total of 14 bumblebees (4 species) visited 75 of these flowers. Each of these bees (as well as several Lepidoptera) extended its proboscis into the flowers as if expecting nectar. The bees were deceived. Apparently they had not yet restricted themselves to specific sites and flowers, and they left the area after visiting an average of only 5.4 flowers. It was not determined whether or not they transferred pollinia. Only 15% of 376 flowers developed seed capsules.

DISCUSSION

The diverse morphological features of flowers attract animals to the nectar and pollen, and increase the *percentage* of pollinating events per visit by the foragers. Has evolution also been operating to force the pollinators to increase the *frequency* of their visits to flowers? Except in special cases, the number of visits to flowers by an individual pollinator is determined by its need for food. It is essential that the pollinators' needs are not too readily satisfied or the attractant signals and flower morphology would be of no value (Heinrich and Raven, 1972). Suitability of a particular animal as an agent of pollen transfer is also related to specificity of movements between flowers and plants of the same species. The specificity, as well as the number of these movements, are functions of the food which is derived from the flowers, the rates at which they can be located, and the ease with which the food can be harvested. Furthermore, the factors affecting the pollination of specific flowers are related to the availability of other flowers in the habitat. An attempt has been made here to quantify information on the energetics of foraging in bumblebees in one habitat where they are the most common pollinators.

The earlier literature on foraging and pollination in bumblebees has been reviewed (Brian, 1954; Medler, 1957). Energetics have received little attention, but is has been concluded that bumblebees learn to forage from the specific flowers from which they have been rewarded (Kugler, 1943; Brian, 1954, 1957; Manning, 1956; Hobbs, Nummi and Virostek, 1961; Hobbs, 1962). The bees may have weak inborn preferences which aid them in initial orientation to flowers, but each bee learns to forage from the flowers

that provide sufficient food and which are best suited to the
bee's body build. It has generally been concluded that the
flower constancy of individaal bumblebees is less than that of
honeybees (Grant, 1950), but Free (1955, 1970) observes that
bumblebees display day-to-day contancy, even when bringing in
mixed pollen loads. He considers that "the inconstancy of bumblebees has been overemphasized." Flexibility in the bumblebees'
foraging behavior is, however, essential in order to exploit new
food resources as they become available throughout the season
(Macior, 1968).

A bee would need to have a large repertory of behavior in
order to forage from all of the diverse flowers available at any
one time. Each flower must be maniuplated in a different way in
order to extract the food. The flowers of *Solanum dulcamara*, for
example, must be shaken by vibrations to release the pollen from
the tubular anthers. The panicles of *Spiraea latifolia* are
"raked" for pollen as the bee twirls around the flower group.
The pollen from *Hypericum esculentum* is swept from the exposed
anthers of individual flowers. In order to collect nectar from
Impatiens biflora the bees must approach and enter a tubular passage of the flowers from a specific direction. The flowers of
Iris versicolor, Rhodora canadensis, Vaccinium sp. and *Uvularia
sessifolia*, and many others, must be approached and entered each
in a specific and different manner (Fig. 4). These specific
movements are learned (Kugler, 1943) and the bees are often initially clumsy and slow. It is unlikely that bees could master
foraging from many different flowers at the same time. The importance of visiting many flowers per unit time in order to make a
profit (Fig. 3) suggests that specialization for foraging to one
flower at a time is energetically desirable.

The vast array of scents, colors and other identifying features of flowers may be energetically important because they
serve to quickly identify the flower to the pollinators. As a
result of these signals the pollinators can presumably locate
specific flowers with a minimum of search. When a forager must
visit two dozen or more flowers per minute, these signals may
save milliseconds in the approach and entrance to the flowers.
But why has evolution produced such a diversity of flower types,
colors and scents rather than just an ideal "bumblebee," "moth"
or other type of flower?

It is obvious that the more alike the identifying signals of
the sympatric and simultaneously blooming species, the more random
will be the transfer of pollen, and the lower the frequency of
pollination in any species. The conditioning of the bees to
specific scents, colors and flower morphology may, however, provide the selective pressure for *divergence* of floral types in sympatric blooming species. The bees will observe their temporary fidelity

Figure 4. Queens of *Bombus vagans* foraging for nectar from flowers of diverse colors and morphology. From top left to bottom right: *Hieracium pratense*, *Iris versicolor*, *Uvularia sessifolia*, and *Rubus strigosus*.

only if the flowers provide sufficient food, and if they can be readily identified. Conversely, there may be advantage to mimicry between rare simultaneously blooming flowers (Macior, 1970).

Tongue length is another factor which affects the frequency and extent to which bees are rewarded. Holm (1966) has reviewed the literature which indicates that bumblebees with long tongues can visit more flowers with a long corolla tube per unit time than the short-tongued bees. Hobbs *et al.* (1961) and Hobbs (1962) tested the foraging preferences of bumblebees on different legumes and concluded that the preferences exhibited by the bees were linked with the ease with which the nectar could be gathered from the flowers. The resulting floral constancy is thus probably explicable in terms of energetics. The bumblebees, as well as honey bees (Butler, Jeffree and Kalmus, 1943; Butler, 1945), forage where they can make the most profits per unit time.

The bees' potential profits are related not only to the food rewards per flower and the morphological features affecting rapid acess to the nectaries and/or pollen. It is related, perhaps more significantly, to the population structure of the flowers. Most of the flowers visited by bumblebees in Franklin County occur in colonies. If the bees were to attempt to forage from more dispersed flowers, their foraging profits would necessarily decrease. It is possible that the dispersed flowers would then become less attractive than those in colonies, and consequently they may produce less seed. Enhanced reproductive success of the plants in colonies may thus, in turn, act to promote patchiness in distribution.

Patchiness of flower distribution can occur in time as well as space. The spacing of flowers can be varied by natural selection on a single plant by blooming synchronously or presenting few blossoms at a time. Synchronous blooming between individuals in the population may similarly affect flower spacing, thus greatly affecting profits of the foragers. Competition by plants for pollinators, as well as competition by pollinators for plants, may exist in the same place at different times of the year (Hocking, 1968; Mosquin, 1971). Macior (1968, 1971) has pointed out the phenological and morphological adaptations between bumblebees and the plants they pollinate. On the one hand it has been suggested that there would be selective pressure for a flower to diverge in blooming from a "cornucopean species" (*Salix* spp. and *Taraxacum officinale* L.) that attracts most of the pollinators (Mosquin, 1971). On the other hand, simultaneous blooming in rare species that are morphologically similar to a more common sympatric species may allow the rarer to be pollinated (Macior, 1970). Presumably if it bloomed alone it would not provide sufficient stimuli to attract the pollinators, who would not find it energetically feasible to forage from the isolated blossoms.

Sparse flower distribution, however, would not necessarily exclude the pollinators if they are amply compensated with sufficient food rewards so that the food intake per unit time is not less. The flowers of *Impatiens biflora*, for example, are produced only a few at a time, and they are sometimes relatively widely dispersed in the foliage. However, presumably because of their high food rewards (Fig. 1) the bumblebees search for them even when they are widely dispersed. Manning (1956) has shown that bumblebees return again and again to a specific site of a plant that provides sufficient nectar. Similar observations were made by Janzen (1971) studying the behavior of Euglossine bees foraging from widely scattered plants in Central America. Presumably those plants which do not regularly provide sufficient food rewards on a regular basis would be deleted during foraging by the bee. When the flowers on the plant in the foraging path of the animal are few, the food reward may have to be more substantial to reinforce the site specificity. However, insects like the *Heliconius* butterflies, which also visit the same plants repeatedly (Gilbert, 1975), have no need to make a rapid energetic profit during foraging, unlike the social bees. This should reduce the necessity for rapid flower visitation, and hence set much different limits for the minimum distribution of the host flowers. Examination of energetic interrelationships of other pollinators and their flowers in various habitats will provide new perspectives and new insights. It is apparent that large contrasts will be found.

SUMMARY

One of the ways of predicting movements of pollinators between flowers and plants is on the basis of energetics. Possible implications of energetics of bumblebees in the pollination of native flora are here examined by way of field observations of foraging, and calculations of approximate energy costs of foraging. Costs of foraging are computed in terms of the food calories in the sugar of the nectar in native host flowers. The observations suggest that most of these flowers must be visited at rapid rates before these pollinators begin to make an energy profit. In order to attract these insects many of the flowers of a species may have to bloom synchronously or grow in colonies. Low temperature is a negligible energy barrier to foraging from some flowers, however it precludes making energy profits from others. The role of flower constancy in relation to energetics and pollination is discussed.

ACKNOWLEDGEMENTS

I thank Dan T. Brackett for assistance in the field and Alan I. Kaplan for help in data processing, also Dr. John Strother for help in identifying some of the flowers. Above all I express my deepest gratitude to Phillip Potter for numerous indirect contributions to the work. Supported in part, by NSF Grant GB-51542.

LITERATURE CITED

Brian, A. D. 1954. The foraging of bumblebees. Bee World 35: 61-91; 81-91.

Brian, A. D. 1957. Differences in the flowers visited by four species of bumble-bees and their causes. J. Anim. Ecol. 26: 71-98.

Butler, C. G., Jeffree, E. P. and Kalmus, H. 1943. The behaviour of a population of honeybees on an artificial and on a natural crop. J. Exp. Biol. 20:65-73.

Butler, C. G. 1945. The influence of various physical and bio- logical factors of the environment on honeybee activity. An examination of the relationship between activity and nectar concentration and abundance. J. Exp. Biol. 21:5-12.

Free, J. B. 1955. The collection of food by bumblebees. Ins. Soc. 2:301-311.

Free, J. B. 1970. The flower constancy of bumblebees. J. Anim. Ecol. 39:395-402.

von Frisch, K. 1967. The dance language of the bees. Harvard University Press: Cambridge.

Gilbert, L. E. 1975. Ecological consequences of a coevolved mutualism between butterflies and plants. This symposium, p. 210.

Grant, V. 1950. The flower constancy of bees. Bot. Rev. 16: 379-398.

Heinrich, B. 1972a. Temperature regulation in the bumblebee *Bombus vagans*: A field study. Science 175:185-187.

Heinrich, B. 1972b. Energetics of temperature regulation and foraging in a bumblebee, *Bombus terricola* Kirby. J. Comp. Physiol. 77:49-64.

Heinrich, B. and Raven, P. H. 1972. Energetics and pollination ecology. Science 176:597-602.

Hobbs, G. A., Nummi, W. O. and Virostek, J. F. 1961. Food-gathering behaviour of honey, bumble, and leaf-cutter bees (Hymenoptera: Apoidea) in Alberta. Can. Ent. 93:409-419.

Hobbs, G. A. 1962. Further studies on the food-gathering be-haviour of bumblebees (Hymenoptera: Apidae). Can. Ent. 94: 538-541.

Hocking, B. 1968. Insect-flower association in the high arctic with special reference to nectar. Oikos 19:359-388.

Holm, S. N. 1966. The utilization and management of bumble bees for red clover and alfalfa seed production. Ann. Rev. Ent. 11:155-182.

Janzen, D. H. 1971. Euglossine bees as long-distance pollinators of tropical plants. Science 171:203-205.

Kugler, H. 1943. Hummeln als Blütenbesucher. Ergeb. Biol. 19: 143-323.

Macior, L. W. 1968. *Bombus* (Hymenoptera, Apidae) queen foraging in relation to vernal pollination in Wisconsin. Ecology 49: 20-25.

Macior, L. W. 1970. The pollination ecology of Pedicularis in Colorado. Amer. J. Bot. 57:716-728.

Macior, L. W. 1971. Co-evolution of plants and animals - syste-matic insights from plant-insect interactions. Taxon. 20: 17-28.

Manning, A. 1956. Some aspects of foraging behavior of bumble-bees. Behaviour 9:164-201.

Medler, J. T. 1957. Bumblebee ecology in relation to the polli-nation of Alfalfa and Red Clover. Ins. Soc. 4:245-252.

Mosquin, T. 1971. Competition for pollinators as a stimulus for the evolution of flowering time. Oikos 22:398-402.

THE ECOLOGY OF COEVOLVED SEED DISPERSAL SYSTEMS

Doyle McKey

Department of Zoology
University of Michigan
Ann Arbor, Michigan 48104

INTRODUCTION

Seed dispersal is a subject that has received curiously lop-sided attention, verging on repetitious overkill from those interested in mechanisms of seed transport but bordering on complete neglect by ecologists. At base, the study of seed dispersal ecology suffers even more from a lack of hypotheses than from a lack of information. Dispersal has been considered the best-known phase of seed and seedling ecology (Stebbins, 1971), but the available information, sometimes considerable, has never been put in order by a unified, systematic set of ideas. The lack of a general theoretical framework has, I believe, hampered collation and extension of the information we already know. The aim of this study is to offer a conceptual basis for the study of the ecology of seed dispersal. Though many of the hypotheses presented cannot be adequately tested with present information, they are offered in the hope they will stimulate further research.

The focus of this study is dispersal as it relates to the life of the individual plant. My hypotheses are designed to answer questions such as the following: what have been the patterns of coevolution between fruits and dispersal agents, and what are the ecological consequences of following different coevolutionary patterns? What adjustments must a plant make to the presence of other plants, potential competitors for dispersal agents? How are the dispersal-based relationships with animal agents and plant competitors organized so that seed dispersal is harmonized with the other elements of a plant's reproductive strategy?

Two major limitations of the study must be mentioned at this point. First of all, though the principles outlined may be somewhat general, most of the examples are drawn from the literature on birds as dispersal agents. Available information on seed dispersal by bats (van der Pijl, 1954) and primates (Hladik and Hladik, 1967) is not extensive. Seed dispersal by rodents, on the other hand, has received much more study, and the coevolution of fruits and rodents (which are usually important seed predators as well as dispersal agents) has been the topic of recent investigations (Smith, 1970; Smythe, 1970) to which this study can add little.

The other major limitation is that while there exists considerable information on how animals exploit fruit crops and on the chemical, morphological, and behavioral features of different fruits and their respective dispersal agents, almost nothing is known of the "seed shadow" (Janzen, 1970) and how selection acts to optimize its form.

PATTERNS OF COEVOLUTION BETWEEN FRUITS AND DISPERSAL AGENTS

A great variety of animals include fruits in their diet, but only relatively few species depend mostly or entirely upon fruits for all the dietary elements. Thus the array of potential dispersal agents varies greatly in the degree to which fruits have influenced their evolution. Fruits are clearly more important bird food items in tropical than in temperate regions. Orians (1969) and Karr (1971) have pointed out the great importance of fruits as a new resource in accounting for the increased diversity of birds in the tropics. Morton (1973) has recently extended this argument to the level of foraging strategies of individual species, pointing out that the availability of fruits in the tropics has led to strategies absent in temperate birds. For example, a very small percentage of tropical birds feed entirely on fruit, even as young. These Morton calls "total frugivores." But much more common is some kind of partial frugivory, usually feeding the young on insects and utilizing fruits heavily as adults.

In this paper I wish to contrast seed dispersal as performed by two different classes of birds differing, in degree and manner, in their dependence on fruits. These I have divided on criteria I feel appropriate for this study. However, they should be roughly equivalent to Morton's total and partial frugivores.

1) Those which eke out all of more of their supplies of carbohydrate, lipid and protein from fruits. The only well-studied birds of this group are the Bearded Bellbird (Snow, 1970) and the Oilbird (Snow, 1962c). Judging from scattered literature (to be presented at appropriate points in the paper), this group also includes some Cotingidae other than the Bearded Bellbird, the Quetzal (*Phasmachrus*) and possibly other Trogonidae, many of most toucans (Ramphasitidae), most of the larger arboreal hornbills (Bucerotidae: Bucerotinae), and probably others. Thus they are restricted to the tropics, but probably generally distributed within forested tropical areas. For the purposes of this paper, such birds will be called "specialized frugivores."

2) Those which utilize fruits primarily as a source of carbohydrates and water. Birds of this group may be primarily insectivorous, even as adults or they may subsist mostly on fruits as adults, catching insects for the young. Whether the

proportion of fruit in the diet of adults is low or high, fruits
are still being exploited in the same manner: as a quickly har-
vestable supply of carbohydrates, water, and possibly minerals
and vitamins that does not much conflict with the harvesting of
proteins and lipids from other sources. Here this will be called
"opportunistic" exploitation of fruits. As the term is used here,
then, probably the majority of neotropical birds are opportunistic feeders on fruits. Many temperate birds also feed opportunistically on fruit.

There is of course no sharp division between these two groups
of birds. In fact, probably many neotropical land birds are
intermediately specialized on fruits. This large group includes
even such reputed frugivores as the manakins (Pipridae), which in
fact take a greater number of insects than do birds herein considered specialized frugivores, and exploit mostly sweet succulent fruits, very different from those exploited by "true"
specialists (Snow, 1971). My concern here is not to draw a
dichotomy but to emphasize what I believe are valid differences
in the interactions with fruits of the two extremes of this continuum - with the knowledge that many birds belong to the undefined
middle portion of the continuum.

Present evidence, due almost entirely to the field studies
of D. W. Snow and B. K. Snow in Trinidad and South America, shows
that the kinds of fruits eaten by specialized frugivores such as
the Oilbird (Snow, 1962c) and the Bearded Bellbird (Snow, 1970)
are as a class very different from the kinds of fruits eaten by
birds which exploit fruit opportunistically. Whereas many fruits
eaten by a wide variety of birds, such as those of *Miconia* spp.
(Land, 1963; Snow, 1966), *Conostegia* sp. (Willis, 1966) (both
Melastomaceae) and *Ficus* spp. (McClure, 1966) (Moraceae), have a
succulent flesh whose main nutrients are probably carbohydrates,
the fruits eaten by these specialized frugivores are characterized
by their firm, dense flesh, rich in fats and proteins. Also,
while most fruits ingested by a variety of birds have small seeds
(such as *Miconia*, *Conostegia* and *Ficus*), the fruits eaten by the
Oilbird and the Bearded Bellbird are characterized, with few
exceptions, by a single rather large seed. Scattered information
on other specialized frugivorous birds indicates that these patterns may be general (Skutch, 1944a,b, 1971; Snow, 1961; Wood,
1924). This information also indicates that these specialized
frugivores are the principal dispersal agents for the large-
seeded nutritious fruits which make up most of their diet and
that these fruits and these birds have undergone "mutual evolution" (B. K. Snow, 1970; D. W. Snow, 1971; Snow and Snow, 1971).
Small-seeded succulent fruits, on the other hand, depend for their
dispersal on a wide variety of birds and mammals. How can these
and other patterns of variation in the dispersal agent-fruit

interaction be explained?

The available evidence, incorporated in the discussion to follow, has led me to believe that the evolution of adaptations to specialized frugivory has resulted in generally increased dispersal quality, and that this difference in dispersal quality is fundamental to the understanding of these different patterns in the coevolution of fruits and dispersal agents.

Under the rubric "quality" I include, for example, reliability of visitation, the probability that an ingested seed will be deposited intact and in a state in which it can germinate readily, and the size of the seeds a dispersal agent finds it profitable to carry. As I will argue in the following section, along all these dimensions the quality of dispersal by specialized frugivores is higher than that performed by birds eating fruits opportunistically. I also include the probability of the seed being dispersed to a favorable site, but so little has this phase of dispersal been studied that no credible predictions can be offered about the relative quality along this dimension of dispersal by specialized frugivores and by opportunistic animals. Under the term "quality" I am not, however, including another important feature of dispersal, capacity of a given dispersal system. This feature I term "quantity." Because there are so many more of them, the subset of animals that eat fruit opportunistically can deliver a much greater *quantity* of dispersal at a greater sustained rate. These differences - the greater quality of dispersal performed by specialized frugivores, the greater quantity of dispersal delivered by opportunistic animals - are central to an appreciation of the coevolution of fruits and dispersal agents.

According to this view, three conditions are of overwhelming importance in shaping the coevolutionary patterns based on dispersal:

1) Dispersal agents are in short supply.

2) The set of potential dispersal agents vary widely in the quality of dispersal they perform, and those performing high-quality dispersal (specialized frugivores) are a relatively small subset.

3) Higher-quality dispersal imposes a higher cost per propagule to the plant, in terms of the more nutritious and expensively-produced fruits required to attract the specialized frugivores.

These three conditions determine that the dispersal component of any viable reproductive strategy must be a compromise. The general patterns of coevolution between fruits and dispersal agents represent the two alternative compromises a plant can make:

1) Evolution of a fruit whose seeds can be dispersed by a great variety of animals, most of which feed on fruit opportunis-

tically and perform relatively low-quality dispersal. In this tactic the shortage of dispersal agents is relieved, but the quality of dispersal is compromised. In a later section I will argue that the evolution of mechanisms for dispersal by inanimate agents, such as wind or the plant itself, are also aimed at relieving the shortage of dispersal agents, and also involve a sacrifice in the quality of dispersal.

2) Evolution of a fruit adapted to a relatively small subset of specialized frugivores performing dispersal of high quality. In this tactic the quantity, or at least the rate, of dispersal is compromised to dispersal quality.

The interesting question of what selective factors determine a plant's evolutionary response at this junction will be taken up later. At this point I wish to characterize these two coevolutionary patterns between plants and dispersal agents.

The quality of dispersal

Reliability of visitation. The quality of dispersal can vary along several dimensions, of which some of the most obvious will be examined here. First of all, we may consider the speed of seed removal and the reliability of visitation by dispersal agents. Since specialized frugivores are more dependent on fruits as food, it can be expected that they will be more attuned to the ripening of fruit crops, so that seed removal will begin soon after seed maturation. It can also be expected that once they locate a desirable fruit crop, they will visit it more faithfully than birds feeding opportunistically on fruits. For example, they will be less likely to abandon a fruit crop if insect abundance undergoes a sudden increase. They should also be less likely to desert a partially-exhausted fruit crop to exploit another crop with newer or more fruits. There is even some evidence that individuals of one specialized frugivore, the Oilbird, are faithful to individual fruiting trees, visiting the same tree on successive nights (Snow, 1961). Taken together, the greater attentiveness and faithfulness of specialized frugivores to the fruit crops they exploit mean that the seeds of these crops should spend the minimum amount of time on the plant following their maturation. This in turn means that 1) the chance they will be killed by predispersal predators is minimized, and 2) the chance that the mature fruits will be allowed to remain on the plant and become rotten, dessicated, or otherwise undesirable to dispersal agents, is also minimized.

Seed treatment within the dispersal agent. The quality of dispersal is also affected by the treatment the seeds receive inside the gut of the dispersal agent. Seeds can be damaged or destroyed by harsh physical and/or chemical treatment during their

passage through an animal, even if the animal is not one adapted
to crushing and digestion of seeds (van der Pijl, 1969). There
is in fact no clear dividing line between seed predators and seed
dispersal agents (Janzen, 1971). We would expect that seeds
adapted for dispersal by ingestion would possess, as part of this
adaptation, protection from the rough chemistry and/or physics
experienced inside the guts of animals. Though quite an array of
such defenses probably exists, the best-known defense is a hard
seed coat that resists grinding and chemical degradation of the
seed (Jenkins, 1969; van der Pijl, 1969).

The interaction between the dispersal agent's gut and the
seed's hard coat has usually not been viewed from quite this perspective. It is often stated that some seeds have "improved
germination" after passing through a vertebrate gut, while, as
Janzen (1971) notes, a more accurate and complete evolutionary
statement would be: Seeds adapted for dispersal including animal
ingestion have evolved hard coats to prevent their being destroyed
by the dispersal agent. When such seeds do not pass through a
vertebrate gut, the unabraded seed coat inhibits germination (cf.
Snow, 1971). In other words, seeds *not* passing through a vertebrate gut suffer *retarded* germination. Following this same line
of reasoning, the usefulness in evolutionary studies of concepts
such as "seed dormancy," if not applied carefully, may also be
called into question. Certain kinds of seed dormancy may be mere
artifacts that are introduced when the experimenter attempts to
study germination of seeds that have been deprived of their normal
interaction with dispersal agents.

Looking more closely from this perspective, we can see that
if the treatment of a given kind of seed in the guts of its set of
dispersal agents is at all variable, then such a mechanism of
seed protection presents two dangers: first if some seeds experience, for some reason, comparatively rough treatment - for
example, if they remain within the gut longer than most seeds -
they may be damaged or destroyed by excessive grinding. Secondly,
if some seeds receive very gentle treatment - for example, if they
are voided more quickly than most seeds - they may be excreted
with so much of the seed coat remaining that germination is still
retarded. As long as there is significant variability in the way
a species' seeds are treated in the guts of its collection of dispersal agents, any given seed hardness represents a compromise,
aimed at minimizing the total seed mortality due to these two
factors.

All the necessary ingredients for production of an interaction of this nature have been shown to exist. In the best study
of this aspect of the dispersal agent-fruit interaction, Rick and
Bowman (1961) noted considerable variability in the treatment of
seeds of the Galapagos Tomato, *Lycopersicon esculentum* var. *minor*,

in the guts of the animal thought to be the principal dispersal agent, the Galapagos Tortoise, *Testudo elephantopus*. For example, in their feeding and seed recovery trials, the first seeds were evacuated 12 days after feeding, and seeds were still appearing in the feces on the 20th day, the last day of collection. Furthermore, they showed that this variability may be an important influence on seed success. Their highest germination percentages were obtained from seeds recovered 12 days after feeding, and germination percentages dropped steadily in batches which had spent more time in the tortoise gut. They concluded that prolonged retention in the tortoise digestive tract might lower seed viability.

The last hypothetical feature of such an interaction - lowered fitness of some seeds due to undertreatment - was not demonstrated in Rick and Bowman's study. This component could account for many of the usually (in this study and others) large proportion of seeds that are voided apparently undamaged, but either fail to germinate or exhibit delayed germination. There is no great conceptual leap from the loss of seed viability due to total lack of treatment by dispersal agents (Snow, 1971) to the as yet undemonstrated loss of viability due to undertreatment.

For no single plant do we have detailed information on how close this picture is to the true nature of the interaction between seed coats and dispersal-agent digestive tracts. We can begin, however, to make predictions concerning the kinds of plants in which seed mortality from overtreatment and from undertreatment (the two factors can be lumped as unpredictability in treatment) will be greatest and in which kinds of plants it will be least. First, if variability in seed treatment is greater between species of dispersal agents than within species, then plants whose seeds are dispersed by a wide variety of animals should suffer a loss in dispersal quality due to these factors greater than that suffered by plants whose seeds are dispersed by a few kinds of animals. Seeds of most species should be subject to even more variable treatment from their collection of dispersal agents than Rick and Bowman (1961) noted for seeds of the Galapagos Tomato in the gut of one kind of animal. Second, it can be argued very convincingly (though at present with no direct support) that seeds adapted for dispersal by specialized frugivores suffer extremely small losses of dispersal quality due to these factors.

The argument is as follows. A feature of many specialized frugivores is the little-muscularized, thin-walled, often small stomach. This feature reflects the relatively small amount of mechanical breakdown necessary for digestion of fruit flesh (Jenkins, 1969). Another feature of many specialized frugivores is the habit of regurgitating larger seeds immediately after the

surrounding flesh has been removed in the stomach, before the
seed can enter and experience the rest of the digestive tract.
Both these features enable the specialized frugivore to exploit a
diet of fruit flesh with minimum time and energy spent in processing seeds (mere ballast to a bird adapted for eating fruit flesh)
and in moving them through the digestive tract. For the plant,
this means that seeds ingested by specialized frugivores should
receive very soft treatment within the gut. The non-muscular
stomach should affect them little physically, and regurgitation
delivers them from traveling the remainder of the digestive tract.
Now, if a plant can evolve a fairly close relationship with
specialized frugivores, it can produce relatively soft seeds and
still expect them to survive their encounter with dispersal agents.
Furthermore, thus permitted to produce soft seeds, it eliminates
the chance that some seeds will suffer fitness loss because of
hard seed coats that were incompletely abraded by the dispersal
agent. In short, treatment of the seed is gentle and, more
importantly, *predictably* gentle. The seed is removed from the
hit-or-miss system of muscular stomachs and the requirement that
some - but not too much - of the seed coat be removed. The plant
can evolve a seed in which features favorable to quick germination and establishment do not have to be compromised to other
features necessary to ensure safe passage through the dispersal
agent.

    If dispersal by specialized frugivores does in fact include
the potential for increased quality along this dimension of dispersal, we are then led to the following prediction: to the
degree that hard seed coats are a defense against the gut of the
dispersal agent (and not against dessication, predators, etc.),
seeds adapted for dispersal by specialized frugivores should have
evolved relatively soft seeds. Available information is neither
systematic nor comprehensive, but it tends to support this prediction strongly. Such a pattern of coevolution could explain why
the seeds of many tropical Lauraceae, a family that figures
strongly in the diets of a number of specialized fruit-eaters,
are comparatively soft (Snow, 1971), and why these soft, large,
seemingly vulnerable seeds are totally unharmed in the stomachs
of the Oilbird and some cotingids. Both the Oilbird and these
cotingids are completely frugivorous, with relatively non-muscular
digestive tracts and the habit of regurgitating large seeds before
they enter the intestine.

    Striking examples of the coevolution of plants with soft
seeds and animals with digestive tracts giving seeds gentle treatment are provided by some mistletoes (Loranthaceae) and the birds
that are reported to be their most important dispersal agents.
Parasitic mistletoes are the plants most often reported to be
dispersed mainly by a small number of species of highly

specialized fruit-eating birds. It seems pretty well established that in the Oriental and Australasian regions certain species of Dicaeidae (flowerpeckers) are by far the most important agents of dispersal of mistletoes, and that mistletoe fruits comprise a large proportion of the diet of these species (Docters van Leeuwen, 1954; Ridley, 1930). A similar relationship has been reported between mistletoes and certain tanagers (*Euphonia* and *Chlorophonia* spp.) and Silky-Flycatchers (*Ptilogonys* spp.) in the New World tropics (Dickey and van Rossen, 1938; Skutch, 1954; Snow and Snow, 1971; Sutton, 1951). It is extremely interesting that both *Euphonia* spp. (Wetmore, 1914) and specialized dicaeids (Docters van Leeuwen, 1954) have highly anomalous non-muscular stomachs, and that the seeds of Loranthaceae have *no* seed coat (Lawrence, 1951; van der Pijl, 1969). The soft mistletoe seeds are to some extent chemically protected from their dispersal agents. The resistant viscin layer surrounding the seed is impregnated with chemicals which hasten the seed's journey through the bird's digestive tract (Wetmore, 1914). This chemical protection, like the mistletoe's association with gentle-gutted birds, should decrease the chance that the soft seeds will be destroyed in the bird's gut.

A number of specialized Dicaeidae possess an amazing adaptation to frugivory (Docters van Leeuwen, 1954). Because they eat some insects, they have retained a small gizzard for grinding this portion of their food. But the gizzard is reduced to an outpocketing from the main digestive tract. Furthermore, fruits are shunted past it in a most intriguing way. When an insect is ingested, a sphincter just below the mouth of the gizzard pouch is closed, and the insect is sent into the gizzard to be ground. When a mistletoe berry is ingested, the sphincter remains open (and the gizzard mouth closes?), and the fruit passes directly into the intestine. Thus the bird is spared the needless grinding of the fleshy fruit, and the intoxication that would result from digestion of the mistletoe seed (cf. Ridley, 1930), and is enabled to evacuate the seed more quickly to make room for more fruit.

Mistletoe seed dispersal in temperate regions, though less well studied, seems to be effected again mostly by relatively specialized frugivores (Arvey, 1951) such as the Phainopepla (*Phainopepla nitens*, Ptilogonatidae), and waxwings (Bombycillidae) (Gill and Hawksworth, 1961). Their handling of mistletoe seeds is perhaps similar to that of the tropical euphonias and dicaeids. The Phainopepla, for example, defecates the soft seeds of mistletoe intact (Crouch, 1943). Mistletoe seeds are often killed if ingested by birds with guts unspecialized for frugivory (Ridley, 1930).

These are the best examples of this coevolutionary pattern that can be gleaned from studies completed so far. My own obser-

vations in tropical Africa and Central America indicate that
specialized frugivores such as toucans and some of the hornbills
are important in the dispersal of soft-seeded fruits of a variety
of unrelated families, such as Lauraceae, some of the Burseraceae
(*Pachylobus* in West Africa), *Afzelia* in the *Caesalpinioideae*,
Connaraceae, Myristicaceae (*Virola* and toucans in Central America,
Staudtia and *Pychanthus* and hornbills in Africa), and others yet
unidentified. The soft-seeded arillate fruits of *Cupania seeman-*
nii (Sapindaceae) and *Cnestidium rufescens* (Connaraceae) are
among the most important food items of the Keel-Billed Toucan
(*Ramphastos sulphuratus*) on Barro Colorado Island (Skutch, 1971).
In my experience, the most glaring exception to the hypothesized
rule that fruits dispersed mostly by specialized frugivorous
birds have soft seeds is that possibly furnished by the palms.
Many of the fruits of this hard-seeded family figure heavily in
the diets of specialized frugivores such as the Oilbird (Snow,
1962c) and hornbills. However, rodents may also be important in
the dispersal of these same species, so that the edible seeds
require protection from rodent predation in the form of a hard,
thick coat. At any rate, the germination pores of palm seeds may
permit unretarded germination even if the seed is still surrounded
by a hard and thick coat after dispersal.

In summary, I believe there has been extensive coevolution
based on the advantages to specialized frugivores of quick and
gentle seed treatment, and the advantages to specialized fruits
of having soft seeds. This coevolution has resulted in increased
quality of dispersal for the plants involved.

It is not known whether quick and gentle seed treatment
decreases the probability that a seed will be carried any distance
from the plant. Do birds equipped for regurgitation or for rapid
seed passage through the gut deposit seeds closer to the plant
than do less specialized birds? This question cannot be answered
at present. While in the bird selection has favored quick and
gentle seed treatment, the plant may have evolved means to make
seed treatment less quick, but still gentle. For example, the
firm dense flesh of most soft-seeded nutritious fruits may have to
be softened in the warm, wet stomach of the bird before the seed
is separated and regurgitated. The firm flesh may thus function
to increase seed retention time without increasing the hazards
faced by the seed. Apparently there is no information in the
details of the regurgitation process - the time span between ingestion and regurgitation, what happens to the fruit within the
stomach before the seed is regurgitated, and so forth. Until this
information is gathered, the total effect of the seed-regurgitation
process on the quality of dispersal cannot be judged.

Seed size tolerated by the dispersal agent. Large seeds have
a greater store of reserves than small seeds, so that in general a
habitat contains a greater number of safe sites for large seeds
than for small seeds (Janzen, 1969). In some environments, selection favors an increase in seed size - and the corresponding decrease in seed number (Harper, Lovell and Moore, 1970; Janzen,
1969).

When selection favors the production of large seeds, how are
these large seeds to be dispersed? To the degree that an animal
is a dispersal agent rather than a seed predator, the seeds within
the fruits it ingests are mere ballast, nutritionally irrelevant
(Snow, 1971). The presence of this ballast, furthermore, may
imply costs in terms of space occupied by ballast that could be
occupied instead by food, and in terms of energy expended on
transporting and expelling the ballast. These costs to the animal
should increase with seed size. While fruits containing a small
mass of seeds may be profitably ingested by a large range of generalized animals, fruits with large seed mass can be ingested
economically only if the dispersal agent has evolved special means
for getting rid of seeds with minimum energetic cost.

Specialized frugivores have evolved such features, chief among
these is the regurgitation of large seeds before they enter the
intestine. The physiological and behavioral equipment associated
with seed regurgitation is possessed by a number of independently
evolved frugivorous birds, such as the Oilbird (Snow, 1962c),
several cotingids (B. K. Snow, 1961, 1970), the Quetzal (Skutch,
1944b), toucans (Skutch, 1944a), many hornbills (personal observation), and probably many others. In these birds, small seeds are
often voided with the feces, as are most seeds ingested by generalized birds, but large seeds, once the surrounding flesh has been
removed in the stomach, are regurgitated. It is probable that the
evolution of the equipment for regurgitation is essential to the
economic exploitation of large-seeded fruits by dispersal agents -
and thus also essential to the evolution of large-seeded fruits
adapted for ingestion. We can envision the following coevolutionary pattern: as fruit becomes an increasingly important part of a
bird's diet, there is selection for increasingly efficient evacuation of seeds. As the bird's efficiency in voiding seeds increases,
it can afford to ingest larger and larger seeds, provided their
accompanying fruit flesh is nutritious enough to make this worth-
while (cf. Snow, 1971). Thus the plant is permitted to evolve
larger seeds (if this move would increase its fitness) provided it
produces a fruit nutritious enough to attract the specialized birds
capable of profitably ingesting large seeds. That quite large
seeds can be profitably ingested and dispersed by a bird possessing
the equipment for regurgitation is shown by the seed sizes of
fruits commonly eaten by the Oilbird (Snow, 1962c). Virtually all

the fruits eaten are characterized by a rather large seed. A
number of the seeds figured are an inch or more in length or diameter, much larger than the seeds that can be economically
ingested by birds of the same size feeding opportunistically on
fruits. The high nutritive content of these fruits has already
been noted.

Many of the fruits reported to be important in the diets of
other specialized frugivores possessing the regurgitation response
also have seeds that are very large compared to those of fruits
ingested by other birds. For example, in addition to the Oilbird,
several cotingids (B. K. Snow, 1961, 1970), toucans (Skutch, 1944a),
and the Quetzal (Skutch, 1944b) also feed heavily on large-seeded
lauraceous fruits. Fruit pigeons of Oceania (Ridley, 1930) and
toucans (Skutch, 1971) feed extensively on large-seeded, thin-
fleshed myristicaceous fruits. Seven of the eight species considered most important for the Keel-Billed Toucan on Barro
Colorado Island (Skutch, 1971) produce large-seeded fruits. Four
of these belong to families noted for producing nutritious fruits
(Palmae, Burseraceae, Myristicaceae). Two others (*Cupania
seemannii*, Sapindaceae; *Cnestidium rufescens*, Connaraceae) have
oily-textured arils, probably relatively nutritious. Large-seeded
Myristicaceae (*Staudtia*, *Pycnanthus*) are also important hornbill
foods in forests of Cameroun, West Africa (personal observation).
It seems certain there has been extensive coevolution along these
lines, involving increased efficiency of seed evacuation by
specialized birds and increased seed size among plants adapted for
dispersal by these birds. It is significant that specialized
frugivorous birds often have a very wide gape for their size
(Ingram, 1958; Jenkins, 1969; B. K. Snow, 1970; D. W. Snow, 1962a)
and a distensible esophagus (Jenkins, 1969). These adaptations may
reflect their ability to ingest large seeds economically, as well
as their ability to feed on a greater size range of fruits than
opportunistic birds.

In this respect, as in many others, the mistletoes are a
group apart from other plants which have coevolved with relatively
specialized frugivores. Most mistletoes have very small seeds.
How can we explain the conspicuous entry of these plants into the
arena of specialized fruits? One hypothesis that comes to mind is
that their parasitic habit entails special dispersal needs, so that
there has been strong selection for features attracting agents
which meet these needs. One of these special requirements is that
the seeds be glued to a tree branch. The mistletoe can partially
accomplish this by producing a seed surrounded with sticky viscin
(Gill and Hawksworth, 1961) and making this viscin disgestible
only with difficulty. Perhaps to accomplish this fully the seed
also must be dispersed by an animal whose gut is relatively gentle
to seeds, leaving the viscin layer intact. The specialized

mistletoe-feeders, in fact, digest only the outer layer of fruit pulp, leaving the viscin layer intact (Sutton, 1951).

Another special requirement is that mistletoe seeds be able to germinate and establish themselves very quickly, before they can be washed from their perch by rain, or blown off by wind. Wetmore (1914) in fact noted that mistletoes are far more abundant in the dry southern half of Puerto Rico than in the wet northern half, and attributed this difference to the greater rainwash of mistletoe seeds in the north. Quick germination requires a soft seed, which in turn, I have argued, implies coevolution with specialized frugivores.

One of the many intriguing possibilities revealed by the studies of B. K. Snow and D. W. Snow in Trinidad is that there may be a guild of mucilaginous-seeded epiphyte-fruits, including mistletoes, epiphytic aroids such as *Anthurium*, and the epiphytic cactus *Rhipsalis*, all with similar dispersal requirements and similar adaptations to meet these requirements - and all dispersed by the same two bird species (Snow and Snow, 1971).

There is again no sharp dichotomy between specialized frugivores and birds which feed on fruits opportunistically, and the nature of the dispersal performed by the variety of birds that are intermediately specialized on fruits cannot be characterized at present. The seed-regurgitation response extends to species such as manakins (Pipridae) which take much fruit and a few insects (Snow, 1962a); other intermediately specialized species, such as the saltators (Jenkins, 1969), seem to regurgitate seeds only seldom. For both manakins and saltators, most of the fruits eaten have small seeds which are passed through the gut rather than regurgitated.

In many tropical birds which depend on both fruits and insects, the stomach is much smaller and less muscular than in totally insectivorous birds (Jenkins, 1969). In birds such as some pigeons which vary a diet of fruit flesh with seeds, there is a muscular gizzard, and it is likely that many seeds of fleshy fruits as well as dry seeds, are digested (Jenkins, 1969). Among these semi-specialized frugivores, there is probably much variation in the quality of dispersal performed.

The cost of high-quality dispersal

It is theoretically obvious that plants should evolve to minimize the cost of their dispersal, and indeed much has been written about the poor nutritional quality of most fleshy fruits (Jenkins, 1969). However, it is just beginning to be appreciated that fruits vary widely in their nutrient content and that, while many fruits are mostly sugar and water, some fruits contain relatively high concentrations of expensively produced nutrients such as oils and

proteins (Morton, 1973; Snow, 1971). For example, fruit flesh of
Cinnamomum elongatum (Lauraceae) contains 9 per cent (dry weight)
protein and 44 per cent fat. Fruits of *Ocotea wachenheimii* contain 14 per cent protein and 34 per cent fat. *Dacryodes* sp.
(Burseraceae) fruits contain 11 per cent protein and 24 per cent
fat. *Bactris cuesa* (Palmae) fruits are 13 per cent protein and
39 per cent fat (all figures from Snow (1962c) and all based on
dry weights). The detailed nutritional composition of the fat-
and protein-rich fruits, for example, the variety of lipids contained in one fruit, or the amino-acid balance of their proteins,
has not yet been investigated.

Production of fat- and protein-rich fruits must entail a high
cost per propagule to the plant. Although it is appreciated that
the evolution of such relatively expensive fruits is tied up with
the evolution of specialized frugivores (Snow, 1971), the selectional basis of this coevolution has never been fully explored.
In this section I discuss the proposition that a relatively
nutritious fruit must be produced if a plant is to command the
relatively high-quality dispersal services of specialized frugivores. According to this proposition, the production of nutritious
fruits is viewed as the principal cost of high-quality dispersal.

If an animal feeds opportunistically on fruits, its selection
of fruits is not predicated on the requirement that fruits supply
it with a balanced diet. Many birds, for example, feed primarily
on insects and utilize fruits as a source of water and carbohydrates (Morton, 1973). Consequently, plants depending on such
animals for their seed dispersal may be able to attract them with
relatively inexpensive, unnutritious fruits. This does not mean
that opportunistic animals are not trying to select nutritious
fruits. It only means that for opportunistic animals with many
alternate food sources, other measures of the food quality of
fruits, for example the smallness of seeds, may be more important
than nutritional content: they will choose to ingest a succulent,
sugary fruit with small seeds rather than a protein-rich fruit with
a large seed, as long as proteins and lipids can be gotten more
easily from another source. On the other hand, if a bird is
dependent on fruits for all its dietary requirements, then it will
select the most nutritious fruits of the available array, and
adapt to the annoyances, such as large seeds, which may accompany
these nutritious fruits. Since the bird's dependence on fruits
entails features which, according to the hypothesis, should increase
the quality of dispersal for the plants whose fruits it exploits,
there may be competition between plants for its special services.
Only plants which produce relatively nutritious fruits can command
its attention.

Available information on highly specialized frugivorous birds
strongly supports at least one part of this argument, that these

frugivores do in fact depend mostly on very nutritious fruits (Snow, 1971). The fruits eaten by the two totally frugivorous species that have been well studied, the Oilbird (Snow, 1962c) and the Bearded Bellbird (Snow, 1970) are characterized by their high content of fats and proteins. Snow (1962c) argues convincingly that the Oilbirds are selecting the most nutritious fruits of the range available. This means that if a plant is to attract such a bird, which (a) visits fruit crops faithfully; (b) regurgitates soft seeds unharmed and in a state in which they can easily germinate, and (c) will ingest and disperse seeds that are relatively very large - services which an opportunistic fruit-eater does not perform - it must pay for these services with a nutritious and expensively produced fruit.

The fruits of parasitic mistletoes, which, as noted before, are dispersed in the tropics mostly by specialized frugivores such as some Dicaeidae, tanagers of the genera *Euphonia* and *Chlorophonia*, and Silky-Flycatchers, have apparently never been well characterized chemically, but there has been speculation (Wetmore, 1914) that they are relatively nutritious. This speculation is strengthened by the fact that for a number of the mistletoe-feeders, such as the Australian painted honey-eater (*Grantiella picta*) (Hindwood, 1935) and the mistletoe-bird (*Dicaeum hirundinaceum*) (Keast, 1958), mistletoe berries are an important nestling food. In another specialized frugivore, the Bearded Bellbird (Cotingidae), the fruits that were fed to nestlings were those of the range available that were richest in protein (Snow, 1970).

Information on other specialized frugivores and the fruits they eat is less complete, but is in accord with the argument presented above. Three frugivorous cotingids observed by Snow (1961) fed mostly on the nutritious fruits of Lauraceae. Lauraceous fruits also appear important in the diet of the Quetzal (Skutch, 1944b) and some toucans (Skutch, 1944a). Fruits of palms, Burseraceae, and Myristicaceae figure largely in the diet of the Keel-Billed Toucan (Skutch, 1971). The oil-rich (and perhaps protein-rich) fruits of palms, Myristicaceae, Burseraceae, and other families are the most important foods of frugivorous hornbills in Cameroun, West Africa (personal observation). Myristicaceous and lauraceous fruits are important items in the diet of fruit pigeons in Australia and Oceania (Ridley, 1930; Freeland, 1972).

How are opportunistic animals prevented from exploiting these nutritious fruits? The large seeds prevent many unspecialized animals from ingesting them. If, as in mistletoes, the seeds are both soft and toxic, birds whose digestive tract would kill the seed may be prevented from ingesting them because of the resulting intoxication (cf. Ridley, 1930, p. 467). The combination of seed softness and toxicity as a mechanism to prevent ingestion of

nutritious fruits by unspecialized birds should be looked for
elsewhere (e.g., Sapindaceae, Connaraceae).

In general, the kinds of fruits which opportunistic animals
most often ingest do tend to be succulent, with sugars or organic
acids the most important nutrients. Also, they tend to be small-
seeded. Examples of the kind of small-seeded, succulent fruits
exploited and effectively dispersed by a great variety of
opportunistic animals are the genera *Miconia* (Land, 1963; Snow,
1965). *Conostegia* (Willis, 1966), *Leandra* (Leck and Hilty, 1968)
and many other Melastomaceae, and *Cecropia* (Eisenmann, 1961),
Ficus (McClure, 1966) and *Musanga* (personal observation) of the
Moraceae. The most common feeder recorded at a *Miconia trinervia*
was a woodpecker (Land, 1963). Willis (1966) recorded many
insectivores and partial frugivores such as woodpeckers, tyrant
flycatchers, thrushes, and wood warblers among the 28 bird species
he noted feeding on the tiny-seeded succulent berries of a
Conostegia. Small seeds make these fruits exploitable even by
birds which do not possess an efficient mechanism for quick seed
evacuation.

It is interesting that most specialized frugivorous birds are
rather large. This is probably because most of the tropical
nutritious fruits on which specialized frugivores depend have large
seeds which cannot be ingested (or at least not profitably) by
small birds (Snow, 1971). Most of the small birds which take much
fruit, such a manakins (Snow, 1962a, 1962b), also eat many insects.
I would argue that this is largely because the reproductive
strategies of plants producing small-seeded fruits (the size most
often eaten by manakins) generally do not favor the production of
the necessarily expensive fruit flesh that would give dispersal
agents a balanced diet.

Comparison of the energetic qualities of fats and carbohydrates leads to some interesting predictions. Fats are a more
compact source of energy than carbohydrates, since they yield more
than twice as much energy per gram upon catabolism. Carbohydrates,
however, provide a quicker source of energy, because they have to
undergo fewer chemical changes in their conversion to the form in
which energy is utilized in the muscles (Klieber, 1961). Now, the
smaller the bird and the higher its metabolic rate, the more need
it has for quick energy in the form of carbohydrates. Thus while
large birds should favor fat-rich fruits, which are intrinsically
more nutritious, small birds should favor sugar-rich fruits, which
are less nutritious but provide a quicker supply of energy. This
could be an additional reason why small fruits with small seeds
are usually succulent and contain much carbohydrate rather than
fat.

DISPERSAL IN THE COMMUNITY CONTEXT

Dispersal agents are a resource, and like any other they may
be in short supply. As long as not all the seeds produced are
dispersed at the optimum rate by the best kind of dispersal agent
of those utilized by the plant, dispersal agents can be considered
to be in short supply. There are various degrees of shortage,
ranging from a more or less severe shortage in which fruits rot or
become dessicated before they are taken by dispersal agents, to
milder shortage, in which the mature fruits spend a long time on
the plant before being removed (more of them then being destroyed
by predispersal predators than if they had been removed quickly),
or in which all the fruits may be removed quickly, but by animals
performing suboptimal dispersal.

Many ornithologists who have observed birds at tropical
"feeding trees" (Leck, 1969; Leck and Hilty, 1968; Morton, 1973;
Willis, 1966) report that fruits are often superabundant. For the
plants involved, this means that there is often a shortage
(whether mild or severe) of dispersal agents. What are a plant's
evolutionary responses to competition with other plants for the
services of a limited supply of dispersal agents?

One group of alternative responses consists of altering the
relation of the plant to its dispersal agents. For example,
selection may alter a fruit so that it is exploitable by a greater
variety of dispersal agents. Thus the shortage of dispersal
agents is somewhat relieved but, as pointed out in a previous
section, this relief may come at the cost of decreased dispersal
quality. It should be remembered that the mass of seeds that can
be transported is one dimension of dispersal quality, and that this
response carries the requirement that the plant possess relatively
small seeds.

Alternatively, a fruit may be evolved that attracts a specialized group of animals, shared with relatively few other plants as
dispersal agents. Such an evolutionary move is accompanied by a
rise in dispersal quality, but entails the cost of producing more
nutritious fruits. Also, while it may assure more exclusive use
of its dispersal agents, this response lowers the actual number of
animals available for dispersal, so that the quantity (or at least
the rate) of dispersal that can be performed is lowered.

A third response with respect to dispersal agents is to evolve
structures for dispersal by inanimate agents, such as wind, water,
or even the plant itself. This response, which entirely removes
the plant from competition for dispersal agents, has the advantages
that the structures involved are probably very inexpensive and that
even a huge number of seeds can be dispersed in a very short period
of time. The chief disadvantages are that most of the seeds are
probably not dispersed very far from the parent plant, and that

mechanisms for dispersal by wind, water and other inanimate agents
are adaptvie in a more restrictive range of habitats (Stebbins,
1971). Also, wind-dispersal especially is effective only for
relatively small seeds.

A second group of alternative responses involve altering the
relation between a plant's fruiting season and those of other
plants. For example, a plant may evolve so that it ripens fruit
at a time when few other plants using the same dispersal agents
are producing fruit. This selective factor was postulated by
Snow (1965) as responsible for the displaced fruiting seasons of
18 tree species of *Miconia* (Melastomaceae) in Trinidad. This
displacement in fruiting seasons would carry the greatest
advantage for plants such as *Miconia* spp. which depend upon a great
proportion of the bird community for dispersal during their fruit
ing seasons, since such plants are most likely to be sharing any
given species of dispersal agent. Conversely, since fruiting season displacement means that fruiting of each species may have to
be crowded into a shorter time period, there could be selection
for utilizing a broader collection of dispersal agents, so that
dispersal is still adequate in the short period available for
fruiting.

The alternative to fruiting season displacement is to spread
fruit production over a long period of time, so that at each point
in time less strenuous demands are being made on the community of
dispersal agents. Since a plant following this tactic can present
only relatively few mature fruits at one time, the plant's ability
to attract dispersal agents may be hampered, especially if the
dispersal agents are opportunistic. This consideration, and the
fact that spreading of the fruiting season causes extensive overlap with the fruiting seasons of other plants, demand that plants
following this tactic have their own reliable subset of dispersal
agents, shared with relatively few other plants.

The total response of a plant to competition for dispersal
agents is a composite of its responses with respect to (1) dispersal agents and (2) timing of fruiting. We turn now to the following question: How are a plant's interaction with dispersal agents
and its interaction with competing plants harmonized to produce a
strategy of dispersal? On the basis of the above discussion, the
following predictions can be made:

1. In plants which produce a great mass of small seeds,
there is selection for a fruit that is exploitable by a wide
variety of animals, so that there are enough agents for dispersal
of the crop. For example, in colonizers of the genus *Miconia* and
stranglers of the genus *Ficus* the great number of small seeds produced can be adequately dispersed only if the fruit is made available to a wide variety of animals. Selection for a broad-niched
fruit is compounded when it is important that seeds be removed

quickly, as well as in great numbers. For example, in plants
which satiate insect seed predators, the need for mass-production
and mass-dispersal of a large seed crop in a very short time (to
maximize seed escape from predators) should have selected for
fruits exploitable by many kinds of vertebrate animals.

Since each species with a broad-niched fruit utilizes a large
proportion of the available dispersal agents when it is fruiting,
there should be strong selection for fruiting seasons of such
species to be displaced one from another. In fact, the best
example of displaced fruiting seasons comes from the genus *Miconia*
(Melastomaceae) (Snow, 1965), colonizers whose fruits are
exploited (and whose seeds are effectively dispersed) by a great
variety of birds (Land, 1963, Snow, 1965). I would predict that
in another group of species with very broad-niched fruits, the
genus *Ficus*, sympatric species will often be shown to have displaced fruiting seasons. There are many plants with broad-niched
fruits, however, that have extended rather than highly seasonal
fruiting. These include *Cecropia* spp. and *Musanga cecropioides*
(both Moraceae; personal observation) and some shrubby Central
American *Miconia* (personal observation). To term these
"exceptions" would be presumptuous at the present time.

Frankie *et al*. (1973) contrast the behavior of "seasonal" and
"extended" fruiters. In their study of tree phenology in a tropical wet forest in Costa Rica, extended fruiting occurred significantly more often in the understory than it did in the canopy.
There is, however, no information on the comparative dispersal-
agent niche breadth for seasonal and extended fruiters.

2. In plants which produce a relatively small crop of large
high-quality seeds, there is less selection for increasing the
supply of dispersal agents, and more selection for features such
as nutritious fruits that insure high-quality dispersal of the
seeds by a specialized subset of animals. For the following
reasons, it may be expected that the fruiting season of such a
plant species will often be spread out over a relatively long
period:

a. The number of dispersal agents available is smaller
than for plants producing generalized fruits, and may be easily
overloaded by a mass-ripened crop, even if it is rather small.

b. Fewer species are sharing the specialized frugivores,
so there are fewer fruiting seasons with which each such species
must avoid overlap.

c. Such plants are usually not predator-satiators, so
that intra-crop synchrony in seed production may not be required
for predator escape.

d. Since a specialized frugivore must extract a balanced
diet from the fruits available to it at any one time, its existence
may require that several species of fruits, each providing differ-

ent nutrients, have broadly overlapping fruiting seasons. Selection exerted by the frugivores would favor the spread and overlap
of fruiting seasons of fruits supplying different nutrients (and
perhaps the displacement of fruiting seasons of fruits providing
similar nutrients).

There is little available information useful in evaluating
this prediction. Almost all of it comes from Snow's (1962c) study
of the food of the Oilbird. Though Snow notes that most of the
large-seeded nutritious fruits eaten by the Oilbird have a
definite season, many of these seasons are quite long. The mean
fruiting season of the ten most important lauraceous species, for
example, is between three and four months. Individual trees of
the two Burseraceae important to these birds, *Dacryodes* sp. and
Trattinickia rhoifolia, remain in fruit for "many weeks". Individuals of the two important palms, *Euterpe langloisii* and *Jessenia
oligocarpa*, also have greatly extended fruiting seasons, each tree
containing inflorescences and fruit bunches in various stages of
growth and producing a few mature fruits at a time throughout most
of the year. Thus the dispersal capacity of the limited supply of
Oilbirds is not overloaded, and a variety of nutritious fruits is
available to the birds at any one time.

3. Since the capacity of inanimate dispersal agents is much
greater than that of the limited supply of animate agents, the
evolution of mechanisms for inanimate dispersal will often appear
where there has been evolution of extremely large crops of small
seeds, and especially when it is required that the seed crop be
mass-ripened and quickly dispersed. For example, in plants which
are satiating insect seed predators, the evolution of adaptations
for wind-dispersal (e.g., in *Cordia alliodora* and *Ceiba pentandra*)
and auto-dispersal (e.g., in *Sesbania macrocarpa*) has made possible
the quick and complete dispersal of great masses of seeds, a task
which would overtax the community of animate dispersal agents.
Though I know of no relevant data, it is conceivable that the cost
per propagule of mechanisms for inanimate dispersal may be low
relative to the cost of producing fleshy fruits for animate dispersal. If so, wind-dispersal would in another respect be consistent with the production of large crops of small seeds.

Inanimate dispersal should also have evolved in communities
where the supply of potential animate agents is low and/or unreliable. This factor may help explain the preponderance of wind-
dispersal in temperate-zone trees. On certain soils in Malaysia,
wind-dispersed dipterocarps dominate in rain forest associations of
low productivity. The scarcity of potential vertebrate dispersal
agents in these unproductive forests may have led to the evolution
and dominance of wind-dispersed trees (Janzen, 1973).

DISPERSAL AND THE TOTAL REPRODUCTIVE STRATEGY

The first part of this paper described the forms of interactions of plants and animal dispersal agents. The second part
discussed how these interactions are harmonized with the plant's
interactions with other fruiting plants. In this section I
examine how the resultant strategies of dispersal are integrated
into the plant's total reproductive strategy.

If a plant has a reproductive strategy based on many small
seeds (for example, colonizers, satiators of insect predators,
stranglers) it will not be able to afford the high cost per
propagule of quality dispersal (nutritious fruits) for each of
its many seeds, and will have to settle for the lower-quality,
but less expensive dispersal given by opportunistic animals and
by inanimate agents such as wind. Also, their large crops could
not get adequate dispersal from the very limited supply of high-
quality dispersal agents. Thus we should expect to find that
those plants entrusting dispersal of their seeds mainly to opportunistic animals or to inanimate agents such as wind, and thus
receiving relatively low-quality dispersal, should have "r-
selected" reproductive stragegies based on crops of many small
seeds. This pattern is exemplified by many Melastomaceae and
Moraceae, and includes many wind-dispersed species from a variety
of families, e.g. *Cordia alliodora* (Boraginaceae), *Ceiba pentandra*
(Bombacaceae), many Bignoniaceae, many Meliaceae, and most temperate wind-dispersed trees. Those plants which have increased
the quality of their seed dispersal by having coevolved with
specialized frugivores, should have "K-selected" reproductive
strategies based on crops of relatively few large seeds. The
species of palms, Lauraceae, Burseraceae, Myristicaceae, Sapindaceae, Connaraceae and others previously cited as dispersed mostly
by specialized frugivorous birds all produce crops of relatively
few large seeds.

Thus the form of the relationships forged by a plant with
the community of potential dispersal agents and competitors reflects the plant's response to the quantity-quality of dispersal
dilemma. This response in turn is closely tied to the plant's
decision in other compromise situations involving seeds, such as
seed size, number and toxicity (Harper, Lovell and Moore, 1970;
Janzen, 1969).

The relation between a plant's response in compromise situations involving seeds and those involving flowers is less clear.
For example, one large flower can produce many small seeds, as in
the orchids. Also, many small flowers can be produced and the
ovaries of all but a few aborted, potentially leading to production of a crop of a few large seeds. Stebbins (1971), however,
has brought to light a situation in which different flowering

strategies within the Leguminosae may have determined the response
of different species to bruchid seed predation. He points out
that the Mimosoideae, in which the structure of the inflorescence
is associated with the production of many small flowers, have
most often opted for satiation of the bruchids with large crops
of small seeds, while members of the Papilionoideae, in which
crops of a few large flowers are common, most often produce a few
large seeds that are chemically protected from bruchid attack.

Members of different structural components of the plant community will solve their problems of reproduction in different
ways, and this should be reflected in their dispersal strategies.
For example, in energy-poor environments such as shaded forest
understory, strategies involving low allocation of energy to
reproduction (and high allocation to investment out of which future reproduction can result) are favored. In the more energy-
rich environment of emergents and upper canopy members - while K-
selected strategies will still be common (especially if the seedling is to survive in shade) - strategies involving expenditure
of tremendous amounts of energy on reproduction are also allowed
and may often be favored. Thus I would expect that dispersal
strategies involving high total energy expenditure - 1) dispersal
of a great mass of small seeds by a variety of opportunistic animals or by inanimate agents such as wind, and 2) dispersal of
large crops of large seeds by large rodents (agoutis, pacas) and
seed-storing birds which are predators as well as dispersal
agents and must be satiated - will appear most often among emergents and plants of second growth. Seeds adapted for dispersal
by large rodents seem to be mostly limited to canopy members and
emergents in both temperate (e.g., oaks, *Quercus* sp.) and tropical
forests (e.g., *Coula edulis* [Olacaceae] and Brazil nuts [*Berthol-
lettia*, Lecythidaceae]), as are wind-dispersed seeds in tropical
forests (Frankie *et al.*, 1973; Keay, 1957). The closed nature of
many tropical forests of course precludes wind-dispersal in fruit
crops located beneath the canopy level. Large crops of succulent,
small-seeded fruits are characteristic of second-growth plants
(Morton, 1973) and of some forest emergents such as strangler
figs (*Ficus* spp.).

MIMICRY IN SEED DISPERSAL - AN APPENDIX

In the tropics of both hemispheres there occur a number of
woody legumes with rather large, hard, shining seeds, usually
bright scarlet red or bicolored red and black (see van der Pijl,
1969 for photographs). These are the "imitation arils" of Ridley
(1930) and the "mimetic seeds" of van der Pijl (1969). Such seeds
occur in scattered members of other families (van der Pijl, 1969),
but are best known in Leguminosae, to which this discussion is

limited. Fruits of these plants exhibit characters associated
with adaptation to dispersal by birds, but offer no reward (such
as a fleshy fruit) to potential dispersal agents. They are
brightly colored, usually at least partially red. In many cases
there are accessory visual attractants, such as the yellow pod
valves of *Adenanthera pavonina* and the persistent red corolla of
some *Rhynchosia* (van der Pijl, 1969). In many cases the seeds
cling to the pod for a long time, so that if they are to be dispersed at all during that time they must be removed by animals.
The seeds of *Erythrina herbacea* in south Texas, for example, are
so firmly attached to the pod that no seeds are removed even when
the plant is submerged in a flooding river (personal observation).
The seeds begin to fall from the plant only several months after
they are matured.

Naturalists have long held the idea - but have never documented it with much observation and experiment - that through
their visual resemblance to fleshy fruits these seeds achieve
dispersal by birds, without expending the energy required to produce fleshy fruits. The selective basis for this kind of mimicry
is that the mimic achieves dispersal at much less than its usual
cost.

Extremely little evidence is available for evaluation of
this idea. Ridley (1930) mentions several instances of birds
feeding on these seeds, but most of these reports are of seed-
cracking birds such as parrots, and not birds which are likely to
be effective in dispersal. So far as I know the only experiment
that has been done to test the idea is that described by van der
Pijl (1969). In feeding experiments which he performed in Java
using the red seeds of *Adenanthera pavonina* (Mimosoideae), grani-
vorous birds refused the seeds, but frugivores (barbets) accepted
them, defecating them intact. Jenkins (1969) observed two species
of Saltators (*Saltator maximus* and *S. coerulescens*) and two tanagers (*Thraupis virens* and *Ramphocelus passerinii*) ingesting the
scarlet seeds of *Erythrina costaricensis* on several occasions.
These are birds which eat much fruit and usually defecate or
regurgitate the seeds intact. This is the sum of the direct evidence that these seeds actually deceive fruit-eating birds in
nature. It is known, however, that they have fooled at least one
taxonomist. One species of the genus *Ormosia* (Papilionoideae)
was somehow originally described as having black seeds enveloped
by a fleshy red aril. This description caused some confusion in
the generic taxonomy, especially when a subgenus *Arillaria* was
named, based on this single species. It is now held that the
description was mistaken, and actually based on a species with
bicolored red-and-black seeds (Merrill and Chen, 1943).

Though at this point I can offer no further direct evidence,
there is enough indirect evidence in support of this idea to

Table 1. Mimetic seeds in the Leguminosae*

Genus	Subfamily	Mimetic types represented	References
Erythrina	Papilionoideae (Tribe Erythrinae)	red, red and black	Krukoff, 1969, 1972
Ormosia	Papilionoideae (Tribe Sophoreae)	red, red and black	Merrill and Chen, 1943; van Meeuwen, 1962; Rudd, 1965
Rhyuchosia	Papilionoideae (Tribe Cajaneae)	blue, red and black	van der Pijl, 1969
Abrus	Papilionoideae (Tribe Abreae, related to Vicieae)	red and black	van der Pijl, 1969
Adenanthera	Mimosoideae	red, red and black	Corner, 1940

*This list must be considered incomplete. According to van der Pijl (1969), mimetic seeds also occur in some *Pithecellobium* spp. (Mimosoideae), in *Batesia floribunda* and possibly in *Cassia costata* (both Caesalpinioideae).

justify pursuing it. The visual resemblance of these fruits to fleshy bird-dispersed fruits is more than incidental. Furthermore, mimetic seeds have probably evolved independently at least five times within the Leguminosae (Table 1). Evidently the mimetic-seed phenotype has some adaptive significance, and is not a neutral trait that has appeared by chance in a single phyletic group.

A set of observations that strengthens the mimicry hypothesis is the information I have compiled on the geographic distributions of the two main types of mimetic seeds, the all-red ones and the ones that are bicolored red and black. There is reason to believe that these two types of mimetic seeds are mimicking two different kinds of fleshy fruits. In many kinds of fleshy fruits, the seed is completely enclosed by the fleshy part. For the purpose of this discussion I will call these "berries" (*not* meant in the strict botanical sense). In other kinds of fleshy fruits the seed is only partly enclosed by the fleshy part, and the exposed portion of the seed surface is shiny and colored contrastingly with the fruit flesh (Ridley, 1930). These I will call "arillate seeds," also not meant in the strict botanical sense. The most common combination is a black seed partially enclosed by a red aril (Ridley, 1930). Usually more surface area is red than black (this is also true for the red-and-black mimetic seeds).

     The first kind of fleshy fruit, the monocolored "berry," has a much broader geographic distribution than the bicolored "arillate seeds." While monocolored "berries" are found throughout the tropics and into temperate regions, bicolored "arillate seeds" are entirely tropical. Within the tropics bicolored "arillate seeds" appear to be most common in the less seasonal tropics (cf. Corner, 1949). The reason for the narrower distribution of bicolored "arillate seeds" is unclear. Possibly, since the seed is large relative to the thin [and often oily (Corner, 1949)] aril, their presence begs the existence of specialized frugivores which can profitably ingest large-seeded fruit. These birds are found only in the tropics, especially in the less seasonal tropics.

     If "mimetic seeds" do actually mimic fleshy fruits, then the distribution of the monocolored "false-berries" and the bicolored "false-arillate seeds" should correspond with the distributions of real "berries" and real "arillate seeds." Because the seed-coat color of plants with mimetic seeds is almost always noted in taxonomic treatises on these plants, I have been able to gather information on the distribution of the two seed-color patterns in a good proportion of the legumes reported to possess mimetic seeds. These include the *Ormosia* of America (Rudd, 1956), China and Indochina (Merrill and Chen, 1943), and Malaysia (van Meeuwen, 1962), and the *Erythrina* of America (Krukoff, 1969) and Australasia (Krukoff, 1972). Information on other genera and areas is fragmentary. The available information conforms to the prediction made above.

     1) In America, *Ormosia*, with most species occurring in the wet tropics, has many species with bicolored seeds. *Erythrina*, with many species in the strongly seasonal tropics and some species extending into the subtropical United States, has monocolored seeds, except for two bicolored species in the wet forests of the Lesser Antilles, and at least one species in Central American wet forests. [The Central American bicolored species, *E. rubrinervia* (Jenkins, 1969) is not mentioned by Krukoff (1969) and it is possible his listing neglects other species.]

     2) Within the American *Ormosia*, the proportion of species with bicolored seeds is highest in wetter, less seasonal areas and lowest in drier, more seasonal areas (see Fig. 1).

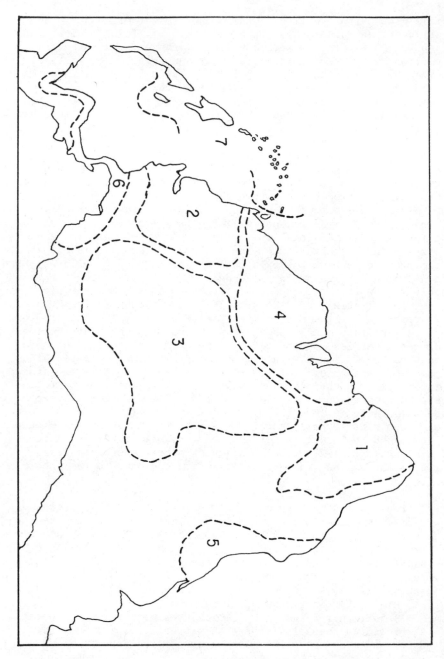

Figure 1. Geographic distribution of seed-coat patterns in American species of *Ormosia*.

3) Most of the Asian and Malaysian *Ormosia* have mono-colored red seeds. The three species with bicolored seeds occur in two of the wettest parts of the Old World range of the genus, the Malay Peninsula (two species) and the Phillipines (one species).

4) Of the two Malaysian species of *Adenanthera* described by Corner (1940), *A. pavonina*, with all-red seeds, occurs from India, southeast China, and Malaysia to the Moluccas, extending into monsoon climates. *A. bicolor*, with bicolored seeds, is restricted to wet forest in Ceylon and Malaya.

5) *Erythrina corallodendron*, with red seeds, occurs in Jamaica and Haiti. *E. corallodendron* var. *bicolor*, with bicolored seeds, occurs in the wetter forests of the lesser Antilles.

Thus bicolored mimetic seeds collectively have a much narrower ecological and geographic distribution than monocolored mimetic seeds, and their distribution patterns correspond to those of real "berries" and real "arillate seeds," lending credence to the idea that the two kinds of mimetic seeds are mimicking different models. Presumably discrimination by birds is the factor that has driven the evolution of red and red-and-black mimics in environments containing high frequencies of the respective types of models. This "fine-tuning" of resemblance to a particular kind of model gains impressiveness when we realize that it has evolved several times. Of the five legume genera in which mimetic seeds have evolved, in three (*Abrus*, with one mimetic species, and *Rhynchosia*, in which monocolored *blue* and bicolored red-and-black seeds have evolved, are the two exceptions) there have been intra-generic switches between red and red-and-black seed types. Within the genus *Ormosia*, red-and-black seededness has probably had at least two independent origins. According to van Meeuwen (1962), the Old World *Ormosia* with bicolored seeds have the black around the hilum, while those in America have the black on the side away from the hilum.

The degree to which birds discriminate will largely determine the effectiveness of the mimetic strategy. Their discrimination should be weakest if they encounter mimetic seeds very infrequently. This consideration leads to the prediction that mimetic seeds will evolve more often in rare species than in common species.

This prediction cannot be adequately evaluated at present. It is interesting to note, however, that many species of *Ormosia* are considered rare by taxonomists who deal with the genus (Merrill and Chen, 1943). As of 1943, 15 of the 34 recognized Chinese and Indochinese *Ormosia* were known from single collections only. Both *Erythrina* and *Ormosia* are rare trees in undisturbed forest in Costa Rica (personal observation, and personal communication, D. H. Janzen).

An intriguing problem is presented by the genus *Ormosia*. In ten American species of this genus, there is great intra-crop variation in seed-coat color. In a single crop, there may be all-red seeds and seeds with varying amounts of black. It is conceivable that this is an adaptive polymorphism: the tree may be producing mimics of each kind of model. By producing two kinds of mimics, the tree lowers the frequency of each mimic relative to that of each model, perhaps increasing the effectiveness of the mimetic strategy. Similar selective forces have produced the well-known polymorphisms in mimetic insects. Van Meeuwen (1962) mentions intra-specific variation in presence or absence of black on the seed of some Malaysian *Ormosia*, but from his discussion it is not clear whether or not the variation is expressed within a single crop, as in the ten American species.

SUMMARY

The major patterns of coevolution between fruits and dispersal agents, as outlined in this paper, can be viewed as the two alternative compromises open to a plant in its dispersal strategy. Whether a plant will opt for a strategy that gives high-quality dispersal but is severely limited in the number of propagules it can accomodate, or for a strategy that insures quick dispersal of a huge mass of seeds, but at the cost of decreased quality of dispersal for each seed, is related to all the other elements of the plant's reproductive strategy: whether it satiates seed predators or produces toxic seeds that confer some freedom from predation, whether it produces small seeds and colonizes light gaps or produces large seeds which can grow in shade, and so forth. In general, species with r-selected reproductive strategies will favor systems such as wind-dispersal and dispersal by opportunistic animals which maximize the number of seeds that can be dispersed and minimize the per-propagule cost of dispersal. Species with K-selected reproductive strategies will favor dispersal by specialized frugivores, so that reliable dispersal is insured for each of the necessarily few expensive propagules.

The form of a plant's interaction with dispersal agents influences how it will interact with potential competitors for these agents. Since each species with a broad-niched fruit (one adapted for exploitation by a great range of animals) utilizes a large proportion of the available dispersal agents when it is fruiting, there might be strong selection for fruiting seasons of such species to be displaced one from another. Conversely, in a species whose fruiting season is squeezed between those of other species, there is selection for a broad-niched fruit so that adequate dispersal can still be obtained in the short time period available for fruiting.

For several reasons, it is expected that species with fruits adapted for exploitation by specialized frugivores will often have long fruiting seasons. 1) The limited number of dispersal agents may be easily overloaded by a mass-ripened fruit crop. 2) Fewer species are sharing the specialized frugivores, so there are fewer fruiting seasons with which each such species must avoid overlap. 3) Intra-crop synchrony of fruit production may not be required for seed predator escape, since these plants are usually not predator-satiators. 4) Since a specialized frugivore must extract a balanced diet from the fruits available to at at any one time, its existence may require that several species of fruits, each providing different nutrients, have broadly over-lapping fruiting seasons.

Mimetic systems in seed dispersal have never been closely examined, but appear to be highly evolved. The preliminary dis-cussion of mimicry in seed dispersal included in this paper pre-sents evidence that the two common types of mimetic seeds are mimics of two different kinds of models, and that the distribution of each mimetic type corresponds with that of its models.

ACKNOWLEDGEMENTS

This study was supported by a National Science Foundation Graduate Fellowship. My examination of this area began when I was a student in "Fundamentals" course 71-5 of the Organization for Tropical Studies in Costa Rica. I am grateful to the parti-cipants for making this course such a satisfying and stimulating experience. This manuscript has profited from discussions with S. Bullock, B. Carroll, C. R. Carroll, R. Cherry., W. Freeland, L. Gilbert, D. Gill, D. Janos, D. Janzen, J. Karr, D. Strong, and J. Vandermeer. I thank Susan Abel for typing the manuscript at an incredibly normal time of her life.

LITERATURE CITED

Arvey, M. D. 1951. Phylogeny of the waxwings and allied birds. Univ. Kansas. Publ. Mus. of Nat. Hist. 3:473-530.

Corner, E. J. H. 1940. Wayside trees of Malaya. Govt. Printing Office, Singapore.

Corner, E. J. H. 1949. The durian theory or the origin of the modern tree. Ann. Bot. 13:367-414.

Crouch, J. E. 1943. Distribution and habitat relationships of the Phainopepla. Auk 60:319-333.

Dickey, D. R. and van Rossen, A. J. 1938. The birds of El
    Salvador. Field Museum, Chicago.

Docters van Leeuwen, W. M. 1954. On the biology of some Loran-
    thaceae and the role birds play in their life-history.
    Beaufortia (Amst.) 4:105-208.

Eisenmann, E. 1961. Favorite foods of neotropical birds: flying
    termites and *Cecropia* catkins. Auk 78:636-638.

Frankie, G. W., Baker, H. G. and Opler, P. A. 1973. Comparative
    phenological studies of trees in tropical lowland wet and
    dry forest sites of Costa Rica. J. Ecol. (submitted for
    publication).

Freeland, W. 1972. A rainforest and its rodents. M. Sc. Thesis,
    U. Queensland.

Gill, L. S. and Hawksworth, F. G. 1961. The mistletoes: a lit-
    erature review. U. S. Dept. Agric. Tech. Bull. No. 1242, p.
    1-87.

Harper, J. C., Lovell, P. H. and Moore, K. G. 1970. The shapes
    and sizes of seeds. Ann. Rev. Ecol. Syst. 1:327-356.

Hindwood, K. 1935. The painted honey eater. Emu 34:149-157.

Hladik, C. M. and Hladik, A. 1967. Observations sur le role des
    primates dans la dissemination des vegetaux de la foret
    gavonaise. Biologia Gabonica 3:43-58.

Ingram, C. 1958. Notes on the habits and structure of the
    guacharo *Steatornis caripensis*. Ibis 100:113-119.

Janzen, D. H. 1969. Seed-eaters versus seed size, number, dis-
    persal and toxicity. Evol. 23:1-27.

Janzen, D. H. 1970. Herbivores and the number of tree species
    in tropical forests. Amer. Natur. 104:501-528.

Janzen, D. H. 1971. Seed predation by animals. Ann. Rev. Ecol.
    Syst. 2:465-492.

Janzen, D. H. 1973. Tropical blackwater rivers, animals, and
    mast fruiting by the Dipterocarpaceae. (submitted to Bio-
    tropica).

Jenkins, R. 1969. Ecology of three species of Saltators with special reference to their frugivorous diet. Ph.D. Thesis. Harvard University.

Karr, J. R. 1971. Structure of avian communities in selected Panama and Illinois habitats. Ecol. Monogr. 41:207-233.

Keast, A. 1958. The influence of ecology on variation in the mistletoe-bird (*Dicaeum hirundinaceum*). Emu 58:195-206.

Keay, R. W. J. 1957. Wind-dispersed species in a Nigerian forest. J. Ecol. 45:471-478.

Klieber, M. 1961. The fire of life. Wiley: New York.

Krukoff, B. A. 1969. Supplementary notes on the American species of *Erythrina*. III. Phytologia 19:113-175.

Land, H. C. 1963. A tropical feeding tree. Wilson Bull. 75:199-200.

Lawrence, G. H. M. 1951. Taxonomy of vascular plants. MacMillan: New York.

Leck, C. F. 1969. Observations of birds exploiting a Central American fruit tree. Wilson Bull. 81:264-269.

Leck, C. F. and Hilty, S. 1968. A feeding congragation of local and migratory birds in the mountains of Panama. Bird-banding 39:318.

McClure, H. E. 1966. Flowering, fruiting and animals in the canopy of a tropical rainforest. Malayan Forester 29:182-203.

van Meeuwen,   1962. Preliminary revisions of some genera of Malaysian Papilionaceae IV. A revision of *Ormosia*. Rein-wardtia 6:225-238.

Merrill, E. D. and Chen, L. 1943. The Chinese and Indo-Chinese species of *Ormosia*. Sargentia 3:77-117.

Morton, E. S. 1973. On the evolutionary advantages and disadvantages of fruit eating in tropical birds. Amer. Natur. 107: 8-22.

Orians, G. H. 1969. The number of bird species in some tropical forests. Ecology 50:783-801.

van der Pijl, L. 1957. The dispersal of plants by bats (Cheirop-
    terochory). Acta Bot. Nederlandica 6:291-315.

van der Pijl, L. 1969. Principles of dispersal in higher plants.
    Springer-Verlag: New York.

Rick, C. M. and Bowman, R. I. 1961. Galapagos tomatoes and tor-
    toises. Evol. 15:407-4 7.

Ridley, H. N. 1930. The dispersal of plants throughout the
    world. L. Reeve and Co.: Ashford, Kent.

Rudd, V. E. 1965. The American species of *Ormosia* (Leguminosae).
    Contr. U. S. National Herb. 32:279-384.

Skutch, A. F. 1944a. Life history of the blue-throated toucanet.
    Wilson Bull. 56:133-151.

Skutch, A. F. 1944b. Life history of the Quetzal. Condor 46:
    213-235.

Skutch, A. F. 1954. Life histories of Central American birds.
    Pacific Coast Avifauna #31.

Skutch, A. F. 1971. Life history of the keel-billed toucan.
    Auk 88:381-424.

Smith, C. C. 1970. The coevolution of pine squirrels (*Tamiasci-
    urus*) and conifers. Ecol. Monogr. 40:349-371.

Smythe, N. 1970. Relationships between fruiting seasons and
    seed dispersal methods in a neotropical forest. Amer. Natur.
    104:25-35.

Snow, B. K. 1961. Notes on the behavior of three Cotingidae.
    Auk 78:150-161.

Snow, B. K. 1970. A field study of the Bearded Bellbird in
    Trinidad. Ibis 112:299-329.

Snow, B. K. and Snow, D. W. 1971. The feeding ecology of tana-
    gers and honeycreepers in Trinidad. Auk 88:291-322.

Snow, D. W. 1961. The natural history of the oilbird. I.
    General behavior and breeding habits. Zoologica 46:27-48.

Snow, D. W. 1962a. A field study of the black and white manakin,

Manacus manacus, in Trinidad.   Zoologica 47:65-104.

Snow, D. W.   1962b.   A field study of the Golden-Headed Manakin, *Pipra erythrocephala*, in Trinidad.   Zoologica 47:183-198.

Snow, D. W.   1962c.   The natural history of the Oilbird, *Steatornis caripensis*, in Trinidad, W. I.   II.   Population, breeding ecology and food.   Zoologica 47:199-221.

Snow, D. W.   1965.   A possible selective factor in the evolution of fruiting seasons in tropical forest.   Oikos 15:274-281.

Snow, D. W.   1971.   Evolutionary aspects of fruit-eating by birds.   Ibis 113:194-202.

Stebbins, G. L.   1971.   Adaptive radiation of reproductive characteristics in angiosperms, II.   Seeds and seedlings.   Ann. Rev. Ecol. Syst. 2:237-260.

Sutton, G. M.   1951.   Dispersal of mistletoe by birds.   Wilson Bull. 63:235-237.

Wetmore, A.   1914.   The development of the stomach in the euphonias.   Auk 31:458-461.

Willis, E. O.   1966.   Competitive exclusion in birds of fruiting trees in western Colombia.   Auk 83:479-480.

Wood, C. A.   1924.   The Polynesian fruit pigeon, *Globicera pacifica*, its food and digestive apparatus.   Auk 41:433-438.

TROPICAL FOREST PHENOLOGY
AND POLLINATOR PLANT COEVOLUTION

Gordon W. Frankie

Department of Entomology
Texas A&M University
College Station, Texas   77843

INTRODUCTION

Recent investigations of animal-plant relationships in the New World tropics have attempted to analyze patterns of biological organization (Baker et al., 1975; Emmel and Leck, 1970; Frankie et al., 1974b; Janzen, 1967; Opler et al., 1975; Smythe, 1970; Snow, 1965). These studies demonstrate that even in relatively non-seasonal tropical climates there exists significant structuring.

In this paper, patterns of flowering phenology are presented for wet and dry forest habitats in Costa Rica. Floral resources, which are reflected by the phenology, are categorized according to respective pollination systems, and the organization of the latter is analyzed for trees on a time and spatial basis. Pollination interactions are considered with respect to pollinator breeding cycles and pollinator preference patterns. A preliminary evaluation is offered as to the potential evolutionary impact made by pollinator activities on the organization of floral resources in both ecosystems.

STUDY AREAS AND METHODS

The wet forest site (La Selva) is situated in a relatively non-seasonal lowland rain forest in the Atlantic watershed; the dry forest site (Comelco) is situated in the highly seasonal lowland forest in the Pacific watershed (see map in Frankie et al., 1974b). Information on climate, vegetation and community types is provided by Frankie et al. (1974b) and Holdridge et al. (1971).

Phenological data for both forests are presented in Frankie et al. (1974b) and Opler et al. (1975). Pollination data presented in this paper were gathered from 1968 through 1972. Use of floral syndromes as described by Baker and Hurd (1968), Faegri and van der Pijl (1971), and Percival (1965), together with supplementary observations on flower visitors, permitted categorization of species that reflect interaction with particular types of animal pollinators. Since these categories or pollination systems are built largely on survey rather than experimental data, the listed pollinators should be regarded as the "most probable pollinators."

WET AND DRY FOREST FLOWERING PATTERNS

The general flowering pattern for overstory and understory wet forest tree species is presented in Figure 1. It is of interest that two of the flowering peaks in the overstory appear to be out of phase with two of the understory peaks (Frankie *et al.*, 1974b).

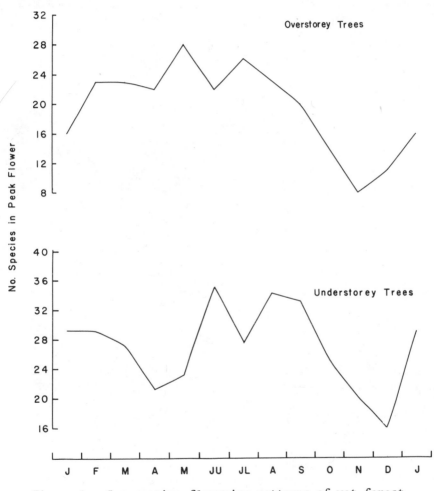

Figure 1. Comparative flowering patterns of wet forest overstory and understory trees in Costa Rica. Patterns based on data compiled by Frankie *et al.* (1974b) for 1969-70.

    Examination of the dry forest flowering pattern also re-
vealed asynchrony. Flowering patterns for dry forest trees and
shrubs are presented in Figure 2. It is of interest that members
of each life form flower largely out of phase with members of
other life forms. Studies by Croat (1969) in Panama reveal similar asynchronous flowering patterns. In the Old World, Boaler
(1966) in Tanzania and D. B. Fanshawe (personal communication) in
Zambia also observed flowering patterns which lack synchronization
among individuals of some represented life forms.

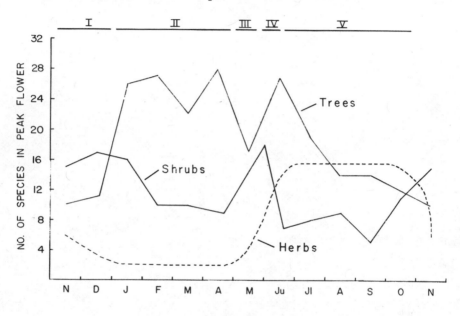

I   Wet season / dry season transition + first part dry season

II Mid dry season

III Late dry season

IV Dry season / wet season transition

V   Wet season

Figure 2. Seasonal flowering patterns of Costa Rican
dry forest plants. Tree pattern based on data in
Frankie *et al.* (1974b), while shrub pattern is based
on information in Opler *et al.* (1975). Herb (tentatively known) and vine patterns are presently being
worked out by Opler, Frankie and Baker.

Adequate explanation for asynchronized flowering in the Costa Rican and Panamanian forests as well as the Old World forests has not been advanced. However, it is important to note that in the dry forest of Costa Rica, this pattern has been observed consistently from 1969-72 and therefore represents a predictable organization of seasonally available floral resources. Regularity in seasonal flowering patterns has also been observed in the Old World. Medway (1972), in his 10-year study of a tropical rain forest in Malaya, documented that the forest community as a whole displayed a marked annual seasonal cycle of flowering.

POLLINATION SYSTEMS OF WET AND DRY FORESTS

Wet forest pollination systems

Pollination systems of wet forest tree species (and other life forms) have received limited attention to date. However, a preliminary survey revealed the following patterns.

Bird pollination. Species believed to be pollinated by hummingbirds are observed in the middle and understory layers of the forest. A few are trees; most are shrubs, vines and epiphytes. Only one canopy species, *Erythrina cochleata*, appears adapted for hummingbird pollination.

Bat pollination. Tree species characteristically adapted for bat pollination occurred primarily in the middle canopy and understory levels. In the wet forest, only two overstory tree species, *Ceiba pentandra* and *Ochroma lagopus*, showed adaptation for bat pollination. Both species are pioneers and grow to assume emergent positions in the canopy.

Diversity of pollination systems. Overstory and understory trees contain a diverse array of pollination systems, however understory trees appear to have a greater representation of each type of system. Support for this observation comes from a preliminary survey in which the most common type of "system" found in the overstory (about 70%) was that of the massively-flowering individual (produces many small flowers, each of which contains a small amount of reward). These trees appear to be adapted for bees, moths, or a combination (opportunistic) of bees, butterflies, wasps and beetles. In addition, most overstory species flower on a seasonal basis (Frankie *et al.*, 1974b). In the understory, fewer species possess the massively-flowering characteristic. Furthermore, many of these species bloom for extended time periods rather than on a seasonal basis (Frankie *et al.*, 1974b).

Evidence for organization on a time basis is limited. Frankie *et al.* (1974b) observed that, in genera containing more than one species, flowering phenologies were segregated seasonally among the respective species of each genus. However, little is known of the pollination systems of these species. Strongest support for temporal organization of floral resources in this forest is found in a study by Stiles (1974). His data suggest that in the case of several sympatric species of *Heliconia*, resources of each species are presented through time as a result of selection by pollinating hummingbirds.

Dry forest pollination systems

Pollination systems of dry forest trees have received considerable attention and therefore will be discussed in more detail. The tree phenology pattern presented in Figure 2 can be divided into "seasonal periods" of flowering based on seasonally-occurring high and low periods of flowering activity. Five principal periods are then recognizeable. Information on pollination systems (based on 70% of the species) operative in each "seasonal period" is presented in Table 1.

By comparing periods I to V (Table 1), three seasonal pollination patterns can be recognized. The most obvious pattern is that of the high frequency of tree species adapted for pollination by medium to large size bees in the middle of the dry season (II). Examples of species include *Andira inermis*, *Myrospermum fructescens* and *Pterocarpus rohrii*. The second pattern is the small bee or opportunistic system which is well represented from one "seasonal period" to the next. *Bursera tomentosa*, *Casearia aculeata* and *Simarouba glauca* exemplify this system. Occurrence of the moth pollination system in mid dry season (II) and in the wet season (IV and V) represents the final pattern. Examples of moth-adapted species include *Cordia alliodora*, *Pithecolobium saman* (dry season) and *Chomelia spinosa*, *Pithecolobium longifolium* (wet season).

Within a "seasonal period," patterns of organization are also apparent. During the wet/dry seasonal transition (I) and the late dry "seasonal period" (III), the small bee or opportunistic pollination system predominates. Corresponding with this observation is the finding of Bawa and Opler (1975) that in addition to the small bee pollination system, most of these species possess a dioecious breeding system. It is noteworthy that the "seasonal period," having the least diverse array of pollination systems, is the late dry period (III). The mid dry period (II) contains, in addition to the high number of medium to large bee-adapted species, the greatest abundance of each represented pollination system. The dry/wet transition (IV) and the wet period (V) are largely represented by the same systems.

Table 1. Seasonal distribution of pollination systems
of dry forest tree species in Costa Rica.

Pollination System[1]	No. Pollination Systems/Seasonal Period				
	I	II	III	IV	V
	Wet/Dry[2]	Mid Dry	Late Dry	Dry/Wet	Wet
Medium-Large Bee[3]	1	20	1	1	3
Small bee or opportunistic[4]	5	9	9	6	13
Moth	0	7	0	3	12
Beetle	2	2	2	3	4
Bat	1	4	0	1	1
Wasp	0	3	0	1	0
Hummingbird	1	1	0	0	1
Butterfly	1	0	0	0	0
Fly	0	1	0	0	0
Wind	0	0	0	1	1

[1]Systems are based on the "most probable pollinator" type.

[2]Includes first 6 weeks of dry season.

[3]Bees belong mostly to Anthophoridae. Lesser numbers are xylocopids.

[4]Most frequent visitors belong to Halictidae, Megachilidae and/or
Meliponini (Apidae). Beetles, flies, wasps and butterflies are
also regular visitors.

Even among tree species adapted to particular pollinator groups (within a given "seasonal period") there is evidence of organization. The 20 tree species serviced by medium to large size bees in the mid dry "seasonal period" (II) flower in a continual sequence from January to April so that overlap among the species populations is minimal (Fig. 3). It has been suggested that because many of these tree species share the same

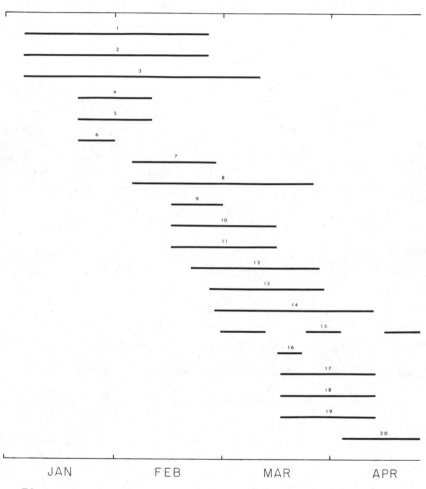

Figure 3. Seasonal progression of flowering phenologies of bee-adapted (medium to large size) tree species in the Costa Rican dry forest. (See appendix for list of tree species 1-20).

anthophilous insects (Opler, unpublished), the plants have been
evolutionarily sorted to allow for reduction in competition for
the pollen vector resource (Frankie *et al.*, 1974a). It is of
great evolutionary consequence to these insects that the flowering sequence is unbroken, thereby providing them with continual
foodstuffs through this period (Baker, 1973).

As part of a biosystematically-oriented investigation, M.
Souza (Universidad de Mexico) has observed a similar phenomenon
in the genus *Lonchocarpus* in central Mexico. He found that each
of seven *Lonchocarpus* species which occur sympatrically in two
different areas (widely separated), flower at different times
and thereby may be avoiding competition for like pollinators.
Snow (1965) documented a similar phenomenon with regard to the
seasonal sequence of fruiting phenologies in several members of
the genus *Miconia* (Melastomaceae) in Trinidad, where fruits are
fed upon by frugivorous birds.

Moth-adapted tree species bloom in a seasonal sequence during
the wet season (IV to V), in a manner which resembles that displayed by the medium to large bee-adapted species of the mid dry
"seasonal period" (II). In contrast to the latter pattern, moth-
adapted species are unevenly distributed through the wet season with
most species flowering during the onset or middle period of the
rainy season (Table 2). Only one species flowered during the
last month of the wet season.

Table 2.   Seasonal distribution of moth-adapted dry
forest tree species in Costa Rica.

"Seasonal period"

	IV	V						I
	Dry/Wet	Wet:	Jul[1]	Jl	A	S	O	Wet/Dry
Number Moth- adapted Spp.	3		5	3	3	4	1	0

[1]Includes most of June

With regard to general structure of pollination systems in
the dry forest, evidence to date indicates these interactions are
organized primarily on a time basis. The highly seasonal climatic
pattern in this forest undoubtedly plays a significant role in this
organization by its influence on the various kinds of pollinator
groups.

RELATIONSHIP OF POLLINATION INTERACTIONS TO POLLINATOR ACTIVITIES

One of the most important aspects of any pollination interaction is that involving the pollinator component. Investigations
of pollinator reproductive cycles and preferences have provided
insight as to possible activities responsible for the evolution and
maintenance of these organized floral resources on a time and
spatial basis.

Reproductive cycles of moths

The hypothesis advanced by Emmel and Leck (1970) for the rain
forest of Panama, in which seasonal fluctuations in butterflies are
associated in part with larval host plant condition, may apply in
part to the seasonal distribution of moth-adapted species in the
Costa Rican dry forest. Moth-pollination systems were observed
during two periods only; the wet season (IV and V) and the mid dry
"seasonal period" (II) (Table 1). In the former period, occurrence of anthophilous moths probably results from populations of
Lepidoptera developing on new foliage resources produced in the
greatest abundance on most plants at this time (see Frankie *et al.*
1974b and Opler *et al.*, 1975, for leafing phenologies). Hawkmoths
(Sphingidae) are known to visit moth-adapted tree species during
the wet season. They have been observed visiting flowers of
Cedrela odorata (onset of rains) and *Caesalpinia coriari* (mid wet
season). Information on seasonal biology of species in this
family and other nocturnal Lepidoptera are not available from this
forest. However, based on observations and collections made at
light in Central America, the first part of the rainy season produces the greatest number of adult moths in a wide variety of
families (T. C. Emmel and J. A. Powell, personal communication).
Studies in Old World tropics tend to support this observation.
Cross and Owen (1970) in Sierra Leone, found that hawkmoths reached
their peak abundance during the onset of the rains.

The number of moth-adapted trees in flower declines in the
latter part of the wet season, and in the subsequent wet-dry transition there are no species in flower which are adapted for moth-
pollination. This pattern may be related to a suggested leaf bud
dormancy operative in at least the trees from August onward
(Frankie *et al.*, 1974b), which may be influenced by the brief dry

spell during July and August. One might speculate that food
resources provided by new leaves in the first part of the wet
season are more suitable for development of most larval populations of moths, and that leaf resources available after this time
are less satisfactory. Therefore, this phase of the reproductive
cycle of most moths may terminate before the wet season is
completed.

     Presence of moth-pollination systems in the mid dry "seasonal
period" (II) (Table 1) may also be related to larval populations
developing at this time. In the dry forest riparian community, 13
of 20 typically riparian species flush large quantities of new
leaves from January to March (Frankie et al., 1974b). This
resource may possibly provide for the development of lepidopterous
pollinators.

     D. M. Windsor (in preparation) completed a study in the Costa
Rican dry forest (Parque Nacional Santa Rosa) that relates to this
possible adult-larval pattern in the Lepidoptera. Through a
monthly census of all phytophagous insects on selected species of
leguminous trees and shrubs, he found two peak periods of caterpillar activity; one at the onset of the rainy season, the other
at the onset of the dry season.

Bee preferences

     In Costa Rica, certain solitary bees have been observed
foraging at the greatest heights in several massively-flowering
tree species (Frankie et al., 1974a). In a preliminary study in
1972, bees were sampled at the greatest and lowest heights throughout the day on one isolated flowering individual of Andira inermis
Collections indicated that the greatest numbers of bees were found
at the greatest height; these were mostly anthophorids and to a
lesser extent xylocopids. At the least height, stingless bees,
megachilids and halictids predominated. This pattern is of
interest with respect to a related investigation involving the
mark-recapture of all flower-visiting bees to A. inermis (Frankie
et al., 1975). Results of this study indicate that several
anthophorid species move at low but consistent rates among individuals of A. inermis, thereby providing a means for outcrossing in
this self-incompatible species.

     In the temperate zone, similar bee preferences have been ob-
served. Levin and Kerster (1973) reported on differences in
height preference among bees visiting tall versus short plants of
Lythrum salicaria in Indiana. In Texas, certain xylocopids show
preference for the highest flowers on vines of Passiflora
incarnata (Frankie, in preparation). Free and Butler (1959, p.
114) report that bumblebees tend to visit flowering trees and
shrubs rather than lower-growing plant forms as light diminishes at
the end of the day.

This bee behavioral pattern may prove to be an important evolutionary pressure in the organization of floral resources at the ecosystem level. Support for this observation is found in an examination of pollination systems operative in the dry forest of Costa Rica during the mid dry "seasonal period" (II). Based on the three life forms[2] represented in Figure 2 and data in Table 1, the medium to large bee pollination system represented in the tree synusia predominates at this time of year. Furthermore, of the few shrubs that flower at this time, most are not adapted for pollination by large bees (Opler, unpublished). In this particular "seasonal period", the combination of large resources at a relatively great height and the seasonal progression of flowering populations appear to be responsible for attraction of most available medium to large size bees to the tree synusia.

Hummingbird preferences

Snow and Snow (1972) investigated feeding habits of nine hummingbird species in the forest of Arima Valley in Trinidad. They divided their birds into two groups, hermits and non-hermits. Members of each group were observed foraging at characteristic levels on certain vegetation types within or adjacent to the forest. However, only understorey plant species were found to be characteristically adapted for hummingbird pollination, in particular by the hermits. This general finding fits well with observations in Costa Rica where plants typically adapted for hummingbird pollination were observed in wet forest understorey vegetation.

In this regard, unpublished studies by F. G. Stiles on hummingbird behavior in California are most informative. Stiles has found that hummingbirds respond in part to the habitat or vegetation containing the flowers. That is, they prefer to forage from the side or from beneath a plant rather than from the top or within the plant. Furthermore, he has found that hummingbirds prefer flowers which face horizontally or which have a downward orientation. Flowers that face upward are not readily visited. If tropical hummingbirds behave similarly to the presentation of flowers and vegetation, then these behavioral preferences may offer some explanation as to why canopy tree species in the Costa Rican wet forest show no apparent adaptation for hummingbird pol-lination.

In addition to foraging patterns of tropical hummingbirds, a simultaneous examination should also be made of activities and interactions with other pollinators that frequent the same canopy layers. It is conceivable that evolutionary pressure of other pollinators may simply be more important at upper versus lower storeys of the wet forest. It is of interest that only one tree

species (*Bourreria quirosii*) in the dry forest (Table 1) shows
adaptation for hummingbird pollination; based on floral behavior,
this system appears to have evolved recently from a moth-
pollinated type.

FLORAL BEHAVIOR PATTERNS VERSUS POLLINATOR BEHAVIOR PATTERNS

Certain patterns of floral behavior may reflect important
pollinator behavior, which then reveals insight into the nature of
pollination systems. For example, extended flowering behavior, in
which relatively few flowers are produced daily over relatively
long time periods, was observed in several wet forest understorey
tree species (Frankie *et al.*, 1974b). Field observations indicate
that respective pollinators display "trap lining" behavior[3]
(Janzen, 1974). That is, these anthophilous animals return daily
to each tree to secure the small amount of reward offered and, in
the process, act as effective outcrossers. This floral behavior
is noticeably absent in overstorey trees in the same forest and is
not characteristic of dry forest tree species.

Other floral behaviors which may reflect important pollinator
behavior patterns undoubtedly exist. For example, it would be of
interest to know why so many wet forest canopy tree species display a massively-flowering behavior. During their flowering
periods, thousands of small, apparently unspecialized flowers
(each with a small amount of reward) are produced. Furthermore,
observations to date indicate that a wide variety of anthophilous
insects are present during the active blooming period of these
species.

EVOLUTION OF POLLINATION INTERACTIONS AT ECOSYSTEM LEVEL

Based on a limited understanding of the possible factors
involved in the organization of pollination systems in the two
Costa Rican ecosystems, it is difficult to advance hypotheses that
would satisfactorily explain how resource structuring has evolved
to its present state in each forest. Historical background for the
evolution of each species undoubtedly differs considerably.
Species of both forests have probably evolved as members (sensu
Baker, 1970) of the respective ecosystems, and information on the
pollinator component of the interactions offers insight into the
evolution.

If we consider the case of a tree species evolving in the dry
forest as a result of invasion from an adjacent ecosystem, the
following developments could conceivably result. It is assumed
that this species has attributes which allow it to be preadapted
abiotically and biotically (with the exception of the pollination
system) to this forest. If the new species possesses one of the

pollination systems operative in the tree synusia of this forest
and also blooms at a time to receive the services of respective
pollinators, then its chances for establishment in this forest are
high. Assuming it becomes established, occurrence of this species
within the respective pollination structure of the ecosystem
undoubtedly will influence other member species. Adjustments in
the form of reciprocating or responsive evolution at the ecosystem
level should then be expected. Segregated flowering phenologies
of species adapted for medium to large bee pollination (Fig. 3)
might in part be the result of reciprocating evolution resulting
from species making their respective adjustments in harmony with
each other.

One might speculate on other kinds of reciprocations that
become operative at the ecosystem level once a sufficient number
of tree species become "members" of a pollination group (either
through invasion or through other means). Using the medium to
large bee pollination system as an example, several reciprocations
seem possible. Firstly, competition for the same pollinators
would probably increase. This is an important problem since most
dry forest tree species are obligated to outcross by virtue of
their self-incompatible breeding system (Bawa, 1974). Competition
would possibly be offset if bee populations correspondingly
increased through time. However, this effect may continue only to
a certain point.

During the sorting out period of like pollination systems at
the ecosystem level, one conceivable reciprocation might be the
development of an alternate-year flowering strategy, which would
have selective value at the specific level. Energetically, the
cost would be less to a tree if it flowered once every two years
(or perhaps three years). Furthermore, since bees in this forest
are known to be low-host specific, they would still be maintained
in the ecosystem provided that enough tree species flowered each
year. However, it is possible that the evolution of alternate-
year flowering has also brought about selection for low-host
specificity in the bees. Another advantage to this strategy is
discouragement of reservoir populations of seed predators. As
Janzen (1971) points out, these predators may play an important
role in the evolution of tropical forests.

If ecological synchronization of an invading species (with
regard to pollination system) is somewhat less than adequate,
compensations of various sorts may be expected. One possible
compensation is a temporal shift in daily presentation of floral
rewards. No tropical examples are available, but one temperate
study suggests this possibility. Linsley et al. (1964) found that
in the Mojave Desert of California, *Oenothera campestris* is widely
distributed and is pollinated mainly by *Andrena boronensis* about
one hour after sunrise. Where *O. campestris* occurs with the more

habitat-restricted *O. kernensis*, flowering time of the former
species appears to have been modified to correspond with that of
O. kernensis. In these situations, *O. campestris* opens its flowers
at sunrise and is pollinated primarily by *A. mohavensis*, which is
the same andrenid bee that pollinated *O. kernensis*.

Compensations in some cases may be extreme and might involve
a shift in pollinator type. One temperate investigation provides
an example. In Texas, Gregory (1963-64) found *Oenothera hartwegii*
to be adapted for pollination by hawkmoths at night. However, in
one area where bees were especially active at *O. hartwegii*, the
flowers opened earlier than elsewhere. This behavior suggests a
local adaptation for pollination services of bees.

Other adjustments, for example, small seasonal shifts in
flowering or extended blooming periods, also seem conceivable.
Since these type of ecotypic differences could be easily measured,
comparative intra-specific pollination studies between adjacent and
widely-separated ecosystems would be extremely enlightening.

If synchronization in the pollination system of an invading
species is less than adequate, marginal existence of the species
might be expected. However, this marginality, which could be
reflected by rareness, may be short-lived. Gentry (personal
communication) observed that *Macfadyena unguis-cati* (Bignoniaceae)
usually flowers at the end of the dry season in the Panamanian dry
forest. However, when it occurs in zones adjacent to or outside
the dry forest, seasonal periodicity becomes irregular. This
results in no fruit set owing to a lack of synchronization in
flowering among members of a population for the outcrossing
services of pollinators. Gentry suggests that this unusual flowering pattern may be a limiting factor in the ecological range of the
plant.

CONCLUSION

It is obvious that our understanding of the underlying pro-
cesses of biological organization of tropical forests (from the
standpoint of plant reproductive biology) is extremely limited and
largely speculative at this time. In structured pollination
systems, careful examination of respective behaviors of pollinators
has provided beginning insight as to the possible source of some of
the selective pressures involved. Behavior is only one aspect of
the pollinator component in pollination interactions; in the future,
corresponding patterns of floral behavior should also be examined.

SUMMARY

Patterns of flowering phenology are presented for wet and dry

forest trees, shrubs and herbs in lowland sites in Costa Rica. In
the case of the tree species of both forests, patterns are broken
down and sorted to specific floral types or systems based on interaction with the most probable pollinator. This categorization allows viewing of the organization of pollination systems on
temporal and spatial bases in the case of wet forest species and on
a temporal basis in the case of dry forest species.

Organized pollination interactions in each forest ecosystem
are considered in light of information on biological and behavioral
features of suspected pollinators. With regard to dry forest trees,
distinct patterns involving bee preferences and moth breeding
cycles are presented and evaluated within the context of this ecosystem. Hypothetical cases, involving invasion by a tree species
from one ecosystem into an adjacent ecosystem with organized pollination systems, are offered.

NOTES

1. A pilot study conducted in the dry forest in Costa Rica during
January 1974 suggests that certain *Centris* species (Anthophoridae)
prefer to forage high in the canopy. Potted shrubs of flowering
Cassia biflora (pollen source) were elevated on wooden towers, the
highest of which was 7 m. Ground-level *Cassia* plants (of equal
flowering intensity) were positioned at the bases of the towers.
Two *Centris* species, which usually forage on the tree *Cochlospermum
vitifolium* (pollen source), visited the potted *Cassias* and consistently preferred plants at the top level.

2. Some vine species, which are adapted for pollination by large
bees, flower at this time; most of these appear to be canopy
species.

3. Free and Butler (1959, p. 123) described a similar foraging
behavior for bumblebees in a temperate agro-ecosystem.

APPENDIX

List of medium to large bee-pollinated tree species in the dry
forest of Costa Rica. Flowering phenologies of the respective
species are presented in Figure 3.

1. *Gliricidia sepium*
2. *Cochlospermum vitifolium*
3. *Caesalpinia eriostachys*
4. *Papilionoideae*
5. *Platymiscium pleiostachyum*
6. *Tabebuia palmeri*

7. *Andira inermis*
8. *Tabebuia rosea*
9. *Caesalpinoideae*
10. *Myrospermum fructescens*
11. *Pterocarpus rohrii*
12. *Cassia grandis*

13. *Guaiacum sanctum*   17. *Dalbergia retusa*
14. *Ximenia americana*   18. *Lonchocarpus costaricensis*
15. *Laetia thamnia*   19. *Piscidia carthagenensis*
16. *Tabebuia neochrysantha*   20. *Lonchocarpus eriocarinalis*

ACKNOWLEDGEMENTS

Research in Costa Rica was supported by the National Science Foundation (research grants GB-7805, GB-25592 and GB-25592A #2). Facilities in Costa Rica were provided by the Organization for Tropical Studies. I am indebted to P. A. Opler for supplying me with observations on pollinators. H. G. Baker, R. R. Fleet, D. H. Janzen and E. L. McWilliams offered constructive comments on the paper.

LITERATURE CITED

Baker, H. G.   1970.   Evolution in the tropics.   Biotropica. 2:101-111.

Baker, H. G.   1973.   Evolutionary relationships between flowering plants and animals in American and African tropical forests. *In* Tropical forest ecosystems in Africa and South America (B. J. Meggers, E. S. Ayensu, and W. D. Duckworth, eds.), pp. 145-159, Smithsonian Institutional Press, Washington, D. C.

Baker, H. G. and Hurd, P. D.   1968.   Intrafloral ecology.   Ann. Rev. Entom.   13:385-414.

Baker, H. G., Frankie, G. W. and Opler, P. A.   1975.   Seed characteristics and seed dispersal systems in relation to forest vegetation types in Costa Rica (in preparation).

Bawa, K. S.   1974.   Breeding systems of tree species of a lowland tropical community.   Evolution 28:85-92.

Bawa, K. S. and Opler, P. A.   1975.   Origin and evolution of dioecism in higher plants (in preparation).

Boaler, S. B.   1966.   Ecology of a miombo site, Lupa North Forest Reserve, Tanzania.   II. Plant communities and seasonal variation in the vegetation.   J. Ecol. 54:465-479.

Croat, T. B.   1969.   Seasonal flowering behavior in central Panama. Ann. Missouri Bot. Gard. 56:295-307.

Cross, R. M. and Owen, D. F.   1970.   Seasonal changes in energy
    content in tropical hawkmoths (Lep., Sphingidae).   Rev. Zool.
    Bot. Afr. 81:109-116.

Emmel, T. C. and Leck, C. F.   1970.   Seasonal changes in organiza-
    tion of tropical rain forest butterfly populations in Panama.
    J. Res. Lepid. 8:133-152.

Faegri, K. and Pijl, van der L., 1971.   The principles of pollina-
    tion ecology.   Pergamon Press, Oxford.

Frankie, G. W., Baker, H. G. and Opler, P. A.   1974a.   Tropical
    plant phenology: applications for studies in community
    ecology.   In Phenology and seasonal modeling (H. Lieth, ed.),
    Springer-Verlag, Berlin (in press).

**Frankie, G. W., Baker, H. G. and Opler, P. A.   1974b.   Comparative
    phenological studies of trees in tropical wet and dry forests
    in the lowlands of Costa Rica.   Jour. Ecol. (in press).**

Frankie, G. W., Opler, P. A., and Bawa, K. S.   1975.   Foraging be-
    havior of solitary bees: implications for outcrossing of a
    neotropical forest tree species (in preparation).

Free, J. B. and Butler, C. G.   1959.   Bumblebees.   Collins, London.

Gregory, D. P.   1963-64.   Hawkmoth pollination in the genus
    Oenothera.   Aliso 5:357-384,385-419.

Holdridge, L. A., Grenke, W. C., Hatheway, W. H., Liang, T. and
    Tosi, Jr., J. A.   1971.   Forest environments in tropical life
    zones, a pilot study.   Pergamon Press, Oxford.

Janzen, D. H.   1967.   Synchronization of sexual reproduction of
    trees within the dry season in Central America.   Evolution
    21:620-637.

Janzen, D. H.   1971.   Seed predation by animals.   Ann. Rev. Ecol.
    Syst. 2:465-492.

Janzen, D. H.   1974.   The deflowering of Central America.   Nat.
    Hist. 83:48-53.

Levin, D. A. and Kerster, M. W.   1973.   Assortative pollination
    for stature in *Lythrum salicaria*.   Evolution 27:144-152.

Linsley, E. G., MacSwain, J. W. and Raven, P. H.   1964.   Compara-

tive behavior of bees and Onagraceae. III. *Oenothera* bees
of the Mojave Desert, California. Univ. Calif. Publ. Entomol.
33:59-98.

Medway, Lord. 1972. Phenology of a tropical rain forest in Malaya.
Biol. J. Linn. Soc. 4:117-146.

Opler, P. A., Frankie, G. W. and Baker, H. G. 1975. Comparative
phenological studies of understorey trees and shrubs in tro-
ical lowland wet and dry forest sites of Costa Rica (in
preparation).

Percival, M. S. 1965. Floral biology. Pergamon Press, Oxford.

Smythe, N. 1970. Relationships between fruiting seasons and seed
dispersal methods in a neotropical forest. Amer. Nat. 104:
25-35.

Snow, B. K. and Snow, D. W. 1972. Feeding niches of hummingbirds
in a Trinidad Valley. J. Anim. Ecol. 41:471-485.

Snow, D. W. 1965. A possible selective factor in the evolution of
fruiting seasons in tropical forest. Oikos 15:274-281

Stiles, F. G. 1974. Ecology, flowering phenology, and hummingbird pollination of some Costa Rican *Heliconia*. Ecology
(in press).

ECOLOGICAL CONSEQUENCES OF A COEVOLVED MUTUALISM BETWEEN BUTTERFLIES AND PLANTS*

Lawrence E. Gilbert

Department of Zoology
University of Texas
Austin, Texas 78712

INTRODUCTION

Animal-plant mutualisms have been extensively described and yet, to date, we have no clear understanding of the general ecological and evolutionary importance of such interactions. Ecologists have practically ignored mutualism when considering the kinds of population interactions which explain community structure and dynamics.[1] Williamson (1972) accurately reflects the tone of recent temperate zone ecology in relegating mutualisms to the level of interesting curiosity.

In this paper, I present a qualitative and empirically derived model of the various order effects of mutualism between pollen and nectar feeding *Heliconius* butterflies and the cucurbit vines, *Anguria*,[2] which they pollinate. The model is based first, on the assertion that mutualism between adult insect and plant will make possible the evolution of increased behavioral capabilities by the insect and second, on the logic that such capabilities will influence the rate, intensity and richness of coevolution between the insect, its prey (in this case *Passiflora*, the larval host), its competitors (mainly other heliconiines), its predators, and its mutualistic associates. In this way the mutualism ultimately will help determine the emergent features of an entire coevolved sub-community.

A linear sequence of words is not the best means of explaining such a highly reticulate ecological system. It is difficult to choose a point of departure since one might begin with a particular group of organisms, with a type of interaction, or with an emergent community property for equally good reasons. My approach is first, to briefly characterize the major groups of organisms involved; second, to explore in detail the interaction between *Heliconius* and *Anguria*, and finally, to hypothesize concerning the ways that this coevolved mutualism has shaped individual, population and community properties of *Anguria*, *Passiflora* and *Heliconius*. The resulting model is summarized in skeletal form in Figure 1 which is perhaps useful to have roughly pictured as one reads the text.

---

*Title in program of the Congress: "Coevolved mutualism between butterflies and plants."

PASSIFLORA: CATERPILLAR FOOD

   Passiflora and several small genera of the Passifloraceae are
the only larval host plants for *Heliconius* butterflies. Most of
the 350+ species recorded for the New World are tropical vines
which display some of the most striking intra- and inter-specific
leaf-shape and stipular variation known for plants (Killip, 1938).
Locally, *Passiflora* species generally exist as low density populations in which individuals are difficult to find without the aid of
egg-laying butterflies. Moreover, because of herbivore damage,
plants are rarely in flower or fruiting condition. More interesting
however, is the pattern of *Passiflora* diversity among many local
habitats. In spite of the vast number of *Passiflora* species
available in the neo-tropics, local habitats always have less than
5% and more typically less than 2-3% of the 350 total.[3] I will pro-
pose causal explanation for both leaf shape variation and apparent
limits to species packing in *Passiflora* in a later section of the
paper.
   As is true of perhaps most plants, *Passiflora* species possess
a range of defensive chemicals which remove all but closely coevolved herbivores from their list of predators and parasites.
Though their chemistry is poorly studied, many *Passiflora* are known
to contain cyanogenic glycosides and/or alkaloids from which
Heliconius probably (but not yet certainly) derive their distasteful
qualities (Brower and Brower, 1964). Different *Passiflora* species
differ in leaf chemistry (they smell different for one thing) and
this is undoubtedly one reason that almost every *Passiflora* species
is unique from the others with respect to the species or combination
of species of *Heliconius* which prefer it for oviposition in a particular area (for example: Alexander, 1961a, Table 1). Other factors may contribute to such host specificity and will be discussed
below.
   The vast majority of *Passiflora* species possess extrafloral
nectar glands on petioles, leaves, stipules, or bracts. These
glands secrete nectar which maintains a defense force of predaceous
ants, vespid wasps and trichogrammatid egg parasitoids. I believe
habitats may differ in the degree to which the resident *Passiflora*
use these different hymenoptera as defense against host specific
herbivores. In a study of *H. ethilla* in Trinidad it was estimated
that over 90% of eggs were killed by parasitoids (Ehrlich and
Gilbert, 1973). In other areas ants are so common on the *Passiflora* (e.g. La Selva, Costa Rica) that it is hard to believe that
they do not account for most of the mortality of eggs and larvae in
spite of the presence of egg parasitoids.
   In addition to mutualistic defense against coevolved herbivores, a few species of *Passiflora* have evolved hooked trichomes
which are highly effective deterrents against some heliconiine
species (Gilbert, 1971a).

Passiflora species differ with respect to pollinators (various bees, hummingbirds) and seed dispersal agents (birds, bats). On the other hand, the most consistent similarity among species of *Passiflora* is the liability that most, if not all, serve as larval host for at least one heliconian.

HELICONIUS: HERBIVORE AND POLLINATOR

Heliconius or "passion flower butterflies" are common and conspicuous features of low to mid-elevation neotropical forests.[4] The larval stages feed on leaves of *Passiflora* from which they presumably sequester the chemical products that make the adult butterfly relatively unpalatable to birds (Brower et al., 1963; Brower and Brower, 1964). Adult *Heliconius* have a reproductive life span of up to six months (Ehrlich and Gilbert, 1973), made possible by their ability to extract nutrients from pollen (Gilbert, 1972).

The behavioral complexity of *Heliconius*:

Several lines of evidence suggest that *Heliconius* may have the most behaviorally sophisticated adult phase among butterflies: 1) Adults roost gregariously (Poulton, 1931; Crane, 1957; Turner, 1971) and individuals are highly faithful to the same roosting spot over extended periods (Benson, 1972). 2) Both adult feeding stations and roosting areas are located by visual navigation under the poor light conditions of early morning (ca. 0530 hr) and late evening (ca. 1800 hr), respectively (Gilbert, 1971b and unpublished data). 3) A circadian rhythm in photic (versus color) sensitivity (S. Swihart, 1963, 1964) is nicely consistent with the needs of visual nagivation early and late in the day. 4) Adult *Heliconius* can be conditioned to discriminate color associated with nectar rewards (C. Swihart, 1971) and preliminary evidence indicates an ability to learn to associate shape with reward.[5] 5) In addition to regular return to roosting areas, *Heliconius* show up at particular nectar and pollen sources with a high degree of temporal regularity from day to day (see Fig. 3 below). This strongly indicates that *Heliconius* possess a kind of circadian memory rhythm akin to that known for bees (Koltermann, 1971). 6) *Heliconius* seem to be uniquely qualified for life in greenhouses, insectaries and all manner of artificial environments.[6] I attribute this to the fact that *Heliconius* orient primarily to learned land marks, one of which, in this case, being enclosure walls. Most other butterflies observed under similar conditions,[7] except when motivated to feed, mate, or oviposit, orient to sunlight and ignore the enclosure walls. Under such circumstances, fatal damage to the wings occurs in just a few days. 7) *Heliconius* have a very broad visual spectrum - possibly the broadest of all animals (Gary Bernard, personal communication). 8) For their body size, *Heliconius* have the largest heads of New World butterflies and probably of all lepidoptera.[8] Given the capability of the *Heliconius* visual system, and given that a butterfly head is mainly compound eyes and optic ganglia, this observation should not be surprising. Indeed, for some other insects, head size and foraging efficiency are known to be positively correlated (Bernstein and Bernstein, 1969). 9) Even in non-visual aspects of behavior, *Heliconius* are highly complex. For instance, female pupae of a number of species in the *hecalasius* and *charitonius* Groups of *Heliconius* (Emsley, 1965) release a pheromone which attracts males, who then sit on the chrysalis waiting for eclosion. This behavior was first noticed by Edwards (1881) in *H. charitonia*.[9] Other species of *Heliconius* appear superficially to act more like typical butterflies in their mate discovery and courtship. But I have gathered circumstantial evidence that in species such as *H. ethilla* which do not utilize pupal mating, males locate the positions of female prepupae by their odor and return daily to the same areas (Gilbert, unpublished data). Moreover, pupae of *Heliconius* make audible squeaks (Alexander, 1961b) and the adults can hear (Swihart, 1967). Thus, even when pheromones are not released by pupae, communication is possible between pupal and adult stages. 10) Insectary and field observations indicate two ways that the visual system of *Heliconius* enhances the more usual chemosensory modes of host plant discovery.[10] First, as females gain experience in oviposition, they begin to associate the shape and form of the host plant with its chemistry. Circumstantial evidence for this is the frequent insectary and field observation of egg-laying females attracted to visually similar non-host plants. Both form perception and the use of shape by egg-laying females is known in *Papilio democleus* L. (Vaidya, 1969a,b).

A second way that *Heliconius* visual sophistication increases the foraging efficiency for larval host involves learning the position of particular *Passiflora* vines and returning on a regular basis. Although such traplining behavior has been demonstrated only with respect to pollen sources (see below) observations of the same marked female routinely visiting the same vine (W. W. Benson, personal communication; Gilbert, unpublished observations) are good indications that *Passiflora* are incorporated into pollen nectar routes. This is almost certainly the case in populations like the Andrew's Trace *H. ethilla* (Ehrlich and Gilbert, 1973) where the spatial aspect of population structure is governed by the distribution of *adult*, rather than *larval* resources (Gilbert, 1971b). 11) Ovipositing females often spend considerable time inspecting the host plant after it has been discovered. It seems reasonable to suggest that this behavior is a visual search for egg predators and for other *Heliconius* eggs or larvae which in many cases are

cannibalistic. The latter hypothesis is supported by the rarity of
two or more independently laid eggs on the same growth point even
under moderate densities of females in the field. (Random oviposition would lead to occasional clumps which are not found.) W. W.
Benson (personal communication) has independently come to similar
conclusions.

Heliconius diversity and host specificity:

    There are about 45 species of *Heliconius* (Emsley, 1965; Brown
and Mielke, 1972), each a specialist on some sub-set of available
passifloraceous species. Local habitats, however, typically contain no more than 10 *Heliconius* species, each associated as a rule
with a different primary host species. There is, therefore, a correspondence between *Passiflora* and *Heliconius* species diversity at
the local level due to a partitioning of host species among the
herbivore species.
    These patterns of host specificity and partitioning in local
Heliconius communities may be the ultimate outcome of chemical
coevolution between the butterflies and *Passiflora* (see Ehrlich
and Raven, 1965). However, indications are that the basis of the
observed pattern is more complex than simple chemical interaction
between plant and coevolved herbivore. For example each of the six
Heliconius species[11] maintained in the tropical insectaries at
Austin has been reared to adulthood on one or more plants not
generally chosen by its ovipositing females and even will feed
upon, and damage, plants which are unsuitable for normal development. Alexander's (1961a) observations are consistent with this
data. Moreover, given the visual component of host location mentioned above, I suspect that non-chemical factors are important in
the evolution of host specificity in *Heliconius* as was found to be
the case in *Euphydryas editha* (Singer, 1971).

Heliconius population biology:

    A two-year population study of *Heliconius ethilla* in Trinidad
revealed remarkable constancy of adult numbers in two adjacent sub-
populations (Ehrlich and Gilbert, 1973). During the entire period,
including two dry seasons, there were no significant changes in the
number of adults estimated at 20 day intervals. Moreover, at no
time was there a difference in size between adjacent sub-populations
occupying similar areas of forest.
    The observed spatial structures of *Heliconius* populations, like
their dynamic aspects, indicate the operation of strong determinis-
tic elements. The research on *H. ethilla* and Turner's (1971) study
on *H. erato* revealed highly sedentary populations which apparently
result from individual home range behavior by adults.

Studies of marked butterflies indicate that the daily reproductive effort (eggs or spermatophores) is very constant and may
even increase as adults approach their maximum age of 6 months
(Gilbert, 1972). The unusually extended reproductive life of
Heliconius can act to buffer perturbations in larval survivorship
(Gilbert, 1971b). For instance, any perturbation in *Passiflora*
availability or in intensity of larval predation which is less than
3-4 months will be buffered. Very few natural disasters such as
complete defoliation of all possible larval hosts by other insects
will last beyond the reproductive life of a female *Heliconius*.

When all these features of population biology, longevity and
reproductive activity are viewed as an interacting system, it is a
simple step to suspect that the pollen often seen on the proboscis
of *Heliconius* has an important nutritive function. Table 1 sum-
marizes the evidence that pollen is an important source of amino
acids for egg production and quite possibly for adult maintenance.

ANGURIA: TRADES POLLEN FOR SEX

Ultimately it is the flowering pattern of their pollen plants
which provide the option of extended adult life and reproductive
effort for *Heliconius*. Although numerous species can be listed as
pollen and/or nectar sources for *Heliconius* (Gilbert, 1972), it is
the little-known, inconspicuous cucurbit genus *Anguria* and a few
species of their relatives, *Gurania*, with which these butterflies
have most conspicuously coevolved. Indeed, I have removed pollen
from museum specimens of *Heliconius* collected from southern Brazil
to Veracruz, Mexico, and have found the characteristic tetrads of
Anguria to be the only pollen consistently represented over this
entire range.

Anguria and *Gurania* are represented by 29 and 73 species, all
restricted to the neotropics (Cogniaux, 1924). Like *Passiflora*,
the species diversity of these cucurbits is strikingly consistent
and low from locality to locality. In each of four rainforest
study sites in Trinidad, Panama and Costa Rica the maximum number
of sympatric, *Heliconius*-visited cucurbits is either 2 or 3, and no
more than 4 even if hummingbird pollinated *Gurania* species are
included. Moreover, as was the case with local *Passiflora* populations, individual plants are widely spaced.

Early systematic treatments of *Anguria* and *Gurania* indicate
that all species are dioecious and that natural populations have
male biased sex ratios (Cogniaux, 1924; Cheesman, 1940). All of the
seven species[12] which I have studied in the field and grown in the
insectary fit this pattern except for *G. levyana*, a hummingbird
pollinated species, which is monoecious.

Anguria flowers are produced in an inflorescence on a long
pedancles if male, in pairs at each node if female. Male flowers

Table 1.   Evidence for pollen feeding in *Heliconius*
(summarized from Gilbert, 1972).

---

1. Distinctive pollen collecting behavior in *Heliconius*.

2. Other species do not accumulate pollen loads when
    visiting *Heliconius* pollen plants.

3. *Heliconius* possess elaborate pollen-processing
    behavior which includes mixing of dry pollen with
    exuded nectar.

4. Morphological details of *Heliconius* proboscis that
    are involved with pollen collecting are lacking in
    non-pollen gatherers.

5. In experiments with artificial flowers, glass beads
    are chosen over sugar water.

6. Active release of amino acids and protein by pollen
    soon after mixing in sugar solution eliminates need
    for chewing or digesting.

7. *Heliconius* assimilate free amino acids and use them
    in egg production.

8. Pollen feeding increases egg production as much as
    5 X over that observed in straight nectar diets.

---

last only one day, then drop off of the inflorescence. Daily production by one inflorescence is typically .3 to .5 flowers per day;
although some species of *Anguria* (e.g. *limonense*) produce 1.0
flowers per day for the first few days of flower production. By
counting the scars of previous flowers and dividing by estimated
daily rate of production, it is possible to estimate the total
time that any particular *Anguria* has had flowers available to pollinators.

Field and insectary observations indicate continuous flowering
by male plants for periods ranging from 6 months to 3 years.
Amazingly enough, a single inflorescence may have a life span of
from 3 months to over a year. A single inflorescence of *A.
grandiflora* Cogn. collected in the Carare Valley, Columbia,
(deposited in the Chicago Field Museum of Natural History) showed
evidence of 410 past flowers and at least 35 buds waiting to

flower when pressed. This represents at least 1 year, probably
more, of continuous pollen and nectar production at one point in
space, as well as life-long feeding stations for any *Heliconius*
fortunate enough to locate this scattered and highly inconspicuous
resource.

But, in spite of such long term predictability in flower
production, there can be great day to day variation in total flower
production on any particular male plant. This fact results from
little or no correlation between the flowering patterns of the different inflorescences on a plant. To illustrate, Table 2 sum-
marizes flower production for 16 inflorescences of an *A. umbrosa*
male for a 16 day period. Notice that daily flower production
varies from 2 to 10 with an average deviation of 1.57 flowers/day
from the mean of 6.06 flowers/day. At the same time, the variation
between inflorescences over the 16 days is much less (an average
deviation of .71 flower/inflorescence from a mean of 6.06 flower/
inflorescence).

Other important variables in the pattern of flower production
include the number of inflorescences per plant (1-20 in *A.
triphylla*; 5-100 in *A. umbrosa*) and the distances between adjacent
flowering male vines (typically from .1 to .5 kilometer). The
implications of such temporal and spatial patterning in resource
availability for understanding the foraging behavior of *Heliconius*
will be discussed in the following sections.

Both temporal and spatial patterns of flower production in
female *Anguria* differ greatly from those of the male vines. Not
only are the female plants less frequent and more widely spaced,
they also cease flowering when a cluster of fruit begins to develop.
There are thus several reasons for the relative rarity of female
flowers in nature.

Mature *Anguria* fruits resemble small cucumbers and each contains 50-60 mature seed. Entire fruits are sometimes found on
the forest floor after being dropped by parrots (beak marks are
distinctive). Squirrels also eat the fruits and may also carry
them away whole since fruits often disappear completely, without a
trace. Green fruits are often destroyed by an unidentified lepidopteran larvae which bores into the fruit.

BUTTERFLY EGGS AND CUCURBIT SEEDS

Mutualisms are those interspecific relationships for which
benefits clearly outweigh costs for each species involved, and for
which net profits to individuals of each species can be translated
into an increase in Darwinian fitness.

DAY

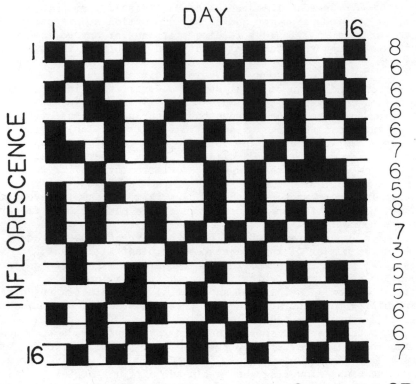

8 5 8 7 4 6 4 6 7 3 10 2 8 6 6 7   97

ANGURIA   UMBROSA

Table 2. Each row represents the flowering of a single
inflorescence of *A. umbrosa* over a 16 day period. Each
column represents a day's flower production by the 16
inflorescences. Solid boxes represent flowers. The
average number of flowers per day and the average number
per inflorescence over 16 days is 6.06.

Benefit versus cost for the butterfly:

The advantage to *Heliconius* of its pollen collecting visits
to *Anguria* are clear (Gilbert, 1972). Studies on the nitrogen
budget of this system show that a large (no. 3 on a scale of 0 to
3) load of *Anguria* pollen contains sufficient nitrogen to account
for the production of 5 eggs of *H. charitonius*,[13] which approxi-
mates the daily egg production of many *Heliconius* species.
What are the costs to *Heliconius*? One kind of cost is the
time and energy required for foraging; another is the increased
risk of predation which is itself a function of time exposed while
foraging. Because *Heliconius* are relatively unpalatable and warningly colored I assume the latter cost to be small.
Although *Anguria* flowers all year-around in rain forests,
with individual male plants flowering constantly for as much as
one year, there is still the previously discussed day to day
uncertainty in the number of flowers available per plant. This
uncertainty in the daily availability of resources at one point
in space is further compounded a) as average number of inflorescences per plant decrease and/or b) as the number competing
Heliconius per inflorescence increase. The greater the uncertainty
per plant, the more plants a *Heliconius* must incorporate into its
trapline to obtain, with high certainty, a given quantity of pollen.
Cost-benefit analysis would predict that whenever the cost of
visiting *Anguria* plants exceeds the benefits of the pollen collected, foraging should ultimately cease. A comparison of female
versus male foraging behavior strongly supports this prediction.
Early morning pollen foragers (ca. 0530-0630 hr) are about 90%
female in Trinidad *H. ethilla* (Gilbert, 1971b). Consequently, in
a total sample of 794 pollen loads, 93% (N = 44) of all *large*
pollen loads were borne by females (Gilbert, 1972) because early
individuals get most of the pollen. Males, for reasons discussed
by Gilbert (1972), benefit less from daily pollen collectings than
do females. One possible exception to this rule occurs on the day
following a mating when males would benefit from rapid replacement
of spermatophore material.[14]

Benefit versus cost for the cucurbit:

Turning now to the plant point of view: what are the benefits
in *Anguria* feeding pollen to *Heliconius*? The obvious answer is
that seed-set requires an animal pollen vector in dioecious plants.
However, if one examines the cost of pollinator service to *Anguria*,
it is seen first, that male plants bear the burden of this responsibility[15] and second, that because females are rare much more
pollen is produced than appears necessary for pollinating the occasional female plants which come into flower. Furthermore, nectar

production, which occurs primarily *after* pollen removal (see Fig.
4 below), is an added cost to the plant which is even more difficult to relate to benefit since the majority of post-pollen nectar
visits come from males not bearing pollen (Gilbert, unpublished
data).

I suggest that the prodigious pollen and nectar production by
male *Anguria* and possibly the evolution of larger, more efficiently
collected pollen grains[16] have evolved by a botanical version of
sexual selection, whereby the *Anguria* male genotype which most consistently keeps *Heliconius* stocked with pollen is the one most
likely to have its pollen land on the rare female stigmas which
sporadically appear throughout the year. In general, any trait of
the male which would enhance the plant's attractiveness to
Heliconius would be selected for the same reason (assuming that
Heliconius can discriminate the different genotypes). Since pollen
collecting by *Heliconius* takes place in semi-darkness for many
Anguria species, the function of subsequent nectar production
during the brighter hours might be to help train the butterflies
to the male plant's position. Nectar production by the female
flower, to be discussed below, may serve a similar function.

Thus, traits of *Anguria* males which might have been hazily
interpreted as "mechanisms to keep the pollinator in the system"
are more plausibly interpreted as traits evolved to maximize
individual fitness.

CONSEQUENCES OF THE *HELICONIUS* X *ANGURIA* MUTUALISM

Pathways by which the *Heliconius* x *Anguria* mutualism
influence both component and emergent properties of a restricted,
coevolved community are shown in Figure 1. The heavy line con-
nects those features of *Heliconius* and *Anguria* most likely
evolved in context of the mutualism: the reproductive longevity
and associated behaviors of *Heliconius* and the flowering pattern
of *Anguria*. Specifically the innovation of pollen feeding by
Heliconius and the constant availability of *Anguria* pollen and
nectar has, by shifting the burden of reproductive effort from
larval to adult stages, increased the reproductive longevity and
reproductive value of adults (Gilbert, 1972). Any physical or
behavioral trait which increases adult foraging efficiency or

Figure 1. Pathways by which *Heliconius* x *Anguria*
mutualism influence individual, population and community level features in a system of interacting
species. See text for explanation.

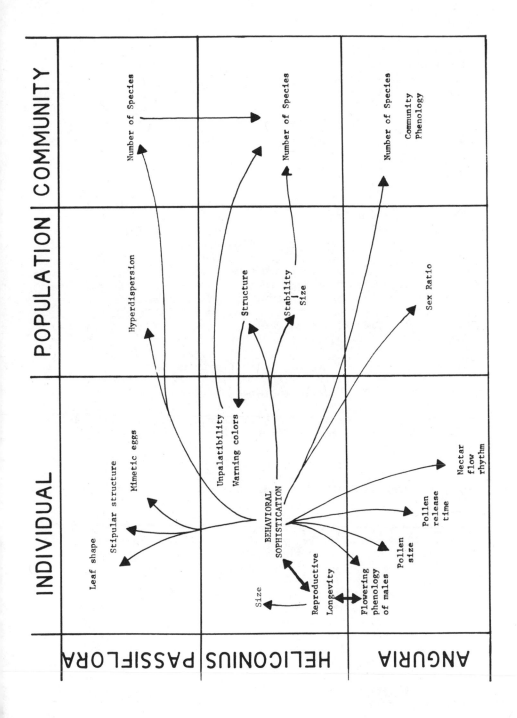

competitive ability will therefore be selected. The highly developed visual system, related head size, learning ability, and early morning flight, etc., are all traits, previously discussed, which can be explained in these terms. Many other properties of the system depend directly on these behavioral capabilities of the butterfly, as will be shown in the following paragraphs.

Navigation, gregarious roosting, kin selection and *Heliconius* diversity:

Since pollen sources are spatially constant over many weeks but inadequate and unpredictable on a daily basis, traplining is clearly an optimal foraging strategy. Perhaps the best evidence for traplining behavior in insects comes from a female *H. ethilla* (#466) whose daily movement patterns can be reconstructed from extensive recapture data Fig. 2). Notice the close correspondence between time and space in the movements of this *H. ethilla* female. Mate-seeking males have been found to behave in similar fashion (Gilbert, 1971b). Such faithfulness to time windows has been described for tropical bees (Janzen, 1971). Although there has been no experimental demonstration of circadian memory rhythm in *Heliconius*, these observations are highly suggestive of such a mechanism. Moreover, the pattern of nectar flow in sympatric *Anguria* is also consistent with this hypothesis (see below).

Roost site fidelity by an individual would improve its ability to locate the first points on traplines under the poor light of early morning. I have hypothesized that gregarious roosting has evolved in *Heliconius* as the result of young individuals following, then roosting, near experienced individuals to enhance their chances of locating scarce, inconspicuous, pollen sources (Gilbert, 1971b). Evidence that experience enhances pollen acquisition is provided in a contrast of pollen load size between young (1 month old) and middle aged (1-3 months old)

Figure 2. Part of the trapline of *H. ethilla* female #466 (from Gilbert, 1971b). On 20 August, 1970, this female was recaptured three times: 1) Station 49, *Gurania spinulosa*, 0550 hr; 2) Station 8, *Palicourea crocea*, 1150 hr; 3) Station 44, *Anguria triphylla*, 1345 hr. On the following day (21 August) #466 was once again seen early at Station 49, and was seen or taken there on two other early morning visits. It was also taken at 49 at 1025 and 1125 hr. The *Anguria* at Station 44 was always visited late by #466: 1410 hr, 1450 hr, and 1345 hr. Stations are 50 yards apart; the line from Station 8 to 12 runs along a ridge; the contour intervals are 100 ft. A map of the entire area is found in Ehrlich and Gilbert (1973).

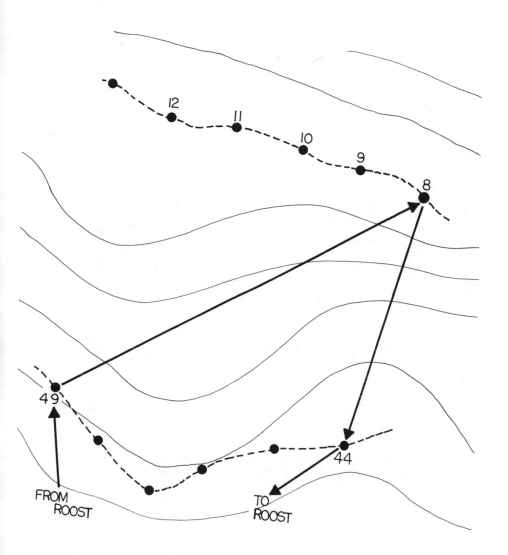

individuals. In a sample of observations from the Trinidad study area, the young and middle aged individuals were equally abundant (40 vs 41%, N = 305) yet among the older individuals large (no. 3) pollen loads were 6X more frequent (Gilbert, unpublished data).

Because individuals are regular in their daily movements, and because of their long potential life span, the butterflies on a roost probably include several generations of closely kin individuals.[17] Benson (1971) and Turner (1971) have suggested that the apparent small size and sedentary nature of many *Heliconius* populations would allow kin selection to operate, accounting for the evolution of unpalatibility and warning coloration in *Heliconius*. Here then is a possible pathway by which *Heliconius* diversity is influenced by increased behavioral sophistication. Richard Levins (1974) has concluded that, in theory, selection for predator avoidance will increase prey populations. Thus, on a given resource base more *Heliconius* individuals can be packed into a habitat because of warning coloration, and more rare species will persist at equilibrium.

Figure 1 indicates other pathways by which increasing behavioral sophistication might determine levels of species diversity for *Heliconius*. Many butterfly species disperse from an area when host plants, nectar sources or other individuals disappear (Gilbert and Singer, 1973). In contrast, because they have the ability to learn the locations of previously visited resources and because of strong roost fidelity, *Heliconius* are less likely to leave an area during temporary shortages of ovipositional sites, mates, etc. This would decrease the probability of local extinction at any given population size and increase the potential number of species at equilibrium. Note that another factor which would enhance *Heliconius* diversity by reducing probability of local extinction was discussed above: namely, the long reproductive life allowed by pollen feeding which acts to buffer population fluctuations.

Circadian isolation and cucurbit species diversity:

Numerous *Heliconius* were observed to visit *Anguria triphylla* in the afternoon and ignore a *Gurania spinulosa* (visited earlier in the morning) clearly visible nearby. These observations stimulated a study of the time course of anther dehiscence and nectar flow in several insectary grown species of *Gurania* and *Anguria* taken from the Trinidad study area.

A striking separation occurs between the sympatric *Anguria* species both in pollen release and nectar flow (Fig. 3). *Gurania spinulosa* (not figured), which releases pollen early with *A. umbrosa*, flows its nectar in the mid day minimizing overlap with the *Anguria* (Gilbert and Golding, in preparation). Whether this temporal displacement is due to selection for increased reproductive isolation or for decreased overlap on the pollinator resource

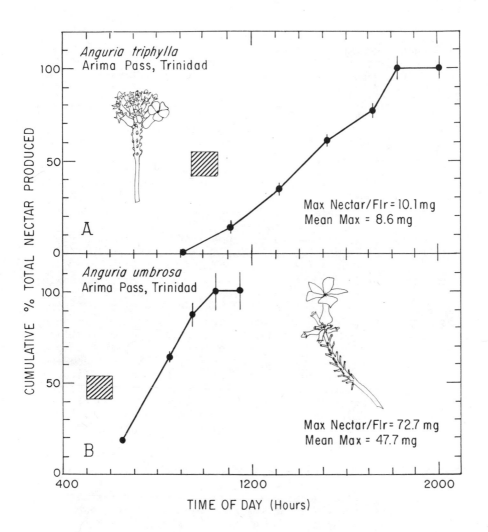

Figure 3. Temporal separation in pollen release (hatched box) and nectar flow patterns in two sympatric *Anguria* species from Arima pass Trinidad. (From Gilbert and Golding, in preparation.) Note that pollen is released before nectar flow is initiated.

is not certain, but the latter possibility is less likely under
limited pollen supply. I know of no other clear example in which
closely related plants with simultaneously opening flowers divide
up a single group of pollinators by evolving different nectar
flow and pollen release rhythms.[18]

The evolution of circadian isolation as is seen in *Anguria* is
most feasible where the pollinator possesses an accurate circadian
memory rhythm. The more finely tuned and sophisticated the memory
rhythms, the more species of closely related sympatric plants
could coexist while utilizing the same pollinator group.
Koltermann's (1971) important work on bees demonstrated that bees
entrained on different scents or colors through a day would,
exactly 24 hours after each entrainment, respond in the appropriate
way as long as different stimuli were originally presented at
intervals of more than 20 minutes.

On the basis of the limited number of sympatric *Anguria* (no
more than two) in all localities so far studied, my guess is that
this interval is much larger for *Heliconius*; i.e., as an isolating
agent to be divided among sympatric *Anguria* species, *Heliconius*
(as a group) can only be divided two or three ways on a daily
basis beyond which isolation would break down; and only future
refinement of *Heliconius* circadian memory will allow more *Anguria*
species to coexist. Reciprocally the plants have apparently
evolved more effective circadian isolation by expanding into the
crepuscular periods at each end of the day. This is indeed possible in the lowland tropics where high nighttime temperatures
allow butterfly flight activity. Indeed, I would hypothesize that
the diurnal shift in *Heliconius* ERG (Swihart, 1963, 1964),
allowing early morning piloting, is ultimately the result of the
butterflies coevolving with the cucurbits in a context of strong
competition for limited pollen resources.

Heliconius learning, sexual selection and sex ration in *Anguria*:

In the previous section I described a way that sexual selection in plants might account for the great amounts of nectar and
pollen given away by *Anguria* in the virtual absence of females.
This hypothesis, which would apply equally well to the male parts
of genetically incompatible or delayed monecious species, depends
upon a pollinator which can discriminate different genotypes of
male plant. In like manner it is possible to account for the
male-biased sex ratio in *Anguria*.

One of the most fascinating population features of both
Anguria and *Gurania* is the fact that sex ratios are heavily biased
in favor of the males (usually > 10:1) in all species across all
areas examined (Trinidad, Panama, Costa Rica; Gilbert, unpublished
data). Indeed, only 20 out of 102 species are described from
female flowers in the only major monograph on these genera

(Cogniauz, 1924). Recently Lloyd (1973) has described similar
male excess in many of the sexually dimorphic perennial Umbelliferae of New Zealand. He accounts for this departure from the
expected 50:50 ratio (Fisher, 1930) by hypothesizing that females
have a shorter life expectancy than males due to greater energy
expenditure in reproduction. Even if valid, this explanation will
not hold for *Anguria* and *Gurania* since the yearly reproductive
effort of a male can easily equal or exceed that of a female.[15]

   Data from sterile plants collected in the field and grown to
flowering and from the few plants that I have managed to grow from
seed to flowering condition indicate either a large male excess in
the primary sex ratio or *extremely* delayed monecious condition.
To date no female plant has appeared among about 10-12 separate
plants of the different species (grown from seed) which have
flowered. While final proof on primary sex ratio is still
pending, I am confident that to think about this sort of model may
be highly useful in interpreting sex ratio data from tropical
plants with intelligent pollinators.

   The model is based upon sex-specific, frequency, and density dependent selection and requires that selection operate at the level of
a sibling cluster. This is possible for plants if seeds drop near
the parent or if dispersed in groups as is the case in *Anguria*
where intact fruits, containing viable seed, are often dropped by
rodents and parrots.

   Anguria female flowers can be considered mimics of the male
flowers. The important signal receiver to be deceived (Wickler,
1968) is the pollen-seeking female butterfly since they collect the
majority of pollen. Deception might occur in several ways: 1)
The female flowering branch in several *Anguria* species is tele-
scoped in such a way as to look like a male inflorescence. 2) The
stigmatic surface is covered with pollen-sized bumps, which may be
more important than flower color mimicry since it is known that
the butterflies confuse pollen and pollen-sized glass beads
(Gilbert, 1972).

   Male-biased sex ratio will occur 1) if *Heliconius* is able to
discriminate and avoid mimics when they become too frequent relative to males in an area, and 2) if those high frequency clusters
of females are the offspring of a female which produces (with high
heritability) equal or female-biased sex ratios in her seedcrops.

   Females in clusters are further disadvantaged by fruit and
seed destruction caused by a moth species whose larvae attack
female but not male reproductive parts. This is an additional sex
specific, density dependent component which could be partly
responsible for evolution of male-biased primary sex ratios in
plants.

Egg-laying *Heliconius*, *Passiflora* leaf-shape diversity, and
egg mimicry:

The apparent importance of vision and visual memory in larval
host plant foraging by egg-laying *Heliconius* stimulated me to
investigate the possibility that these females are agents of
selection on *Passiflora* leaf shape. Under such visual selection
Passiflora would be expected to evolve leaf shapes which would
make them more difficult to locate. There should also be pressure
to diverge from other *Passiflora* species since larval food niche
breadth is considerably broader than that of ovipositing females
and oviposition mistakes occur.[19] The predicted high local
diversity of *Passiflora* leaf shape in contrast to the often noted
monotony of tropical leaves (Richards, 1951) is strikingly verified (Fig. 4). When leaves of two sympatric species are similar
in shape, they usually differ in pubescence, reflectance, stipular
structures or tendrils such that the gestalt of the plants are
different (Gilbert, in preparation).

The literature on leaf shape is concerned almost entirely
with shape as it relates to physical factors. Yet, just as the
first recognition of mimicry in animals was by taxonomists who
simply observed similarities and named species accordingly [e.g.
various genera of moths named Apiformis, Vespiformis, etc.
(Remington, 1963)], so *Passiflora* taxonomists have recognized the
extensive convergence of *Passiflora* leaf shapes on those of common
tropical plants which are essentially vast amounts of inedible
substrate to *Heliconius*. The following species names are found in
Killip (1938): *discoreaefolia*, *morifolia*, *bauhiaifolia*, *tiliae-
folia*, *capparidifolia*, *laurifolia*, *guazumaefolia* and *dalechampi-
oides*. All of these are generic names of common tropical trees
and vines.

There is some evidence from other plants that visual selection may be significant in determining leaf shape. For example,
the striking similarities in both leaf shape and texture between
numerous pairs of species in two south African plant genera,
Cliffortia (Rosaceae) and *Aspalathus* (Fabaceae), (Dahlgren, 1971)
probably reflect strong visual selection from herbivores. The
convergence of Australian mistletoe leaf shapes on the leaf shapes
of their host trees (mostly *Eucalyptus*) (B. A. Barlow and D. Wiens,
personal communication of unpublished data) is in effect the same
phenomenon as was described for *Passiflora* above: (i.e. small
patches of edible plant blending in with a large amount of inedible
foliage).

Further evidence that ovipositing females exert visual selection on *Passiflora* is the presence of butterfly egg mimics on
several species. I first discovered these on *Passiflora cyanea*
(Fig. 5), then on *P. auriculata* (both from Trinidad). Since most

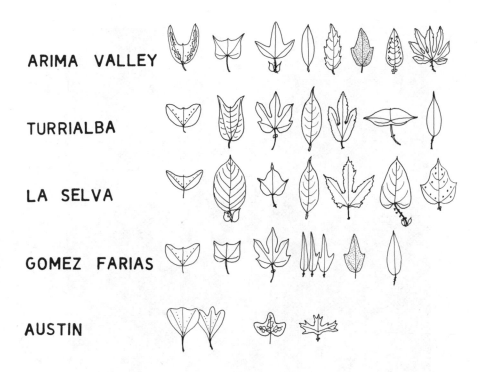

ARIMA VALLEY

TURRIALBA

LA SELVA

GOMEZ FARIAS

AUSTIN

Figure 4. Leaf shape variation among sympatric species of *Passiflora*. The localities are from the top: Trinidad, Costa Rica, Costa Rica, Mexico and Texas. (From Gilbert, in preparation)

Figure 5. Mimetic egg (arrow) on stipule of *Passiflora cyanea*. Several days earlier, new growth point was sheltered within stipule so that fake eggs were presented in the area where eggs are usually deposited. Note extra-flora nectar glands on petiole.

larval *Heliconius* are cannibalistic to a certain degree it makes
sense for the female to detect and reject growth tips of host
which already possess eggs or young larvae. Interestingly the
fake eggs are closer to the color of eggs near the point of
hatching or young larvae (golden) than to the color of newly
oviposited eggs (yellow).[20]

Psychological limits to local *Passiflora* diversity:

    The ability of ovipositing females to learn and discriminate
shapes may greatly influence local *Passiflora* diversity. As was
pointed out above, in spite of some 350 New World *Passiflora*
species, the total number of species packed into a local habitat
rarely exceeds 10. The hypothesis is that the number of *Passiflora*
species should be approximately the number of shape categories
discriminated by *Heliconius*. It can be easily seen that if an
alien *Passiflora* "attempts" to colonize an area in which all possible leaf shapes are already represented it will be discovered by
experienced females searching for its particular leaf category.
Even if chemically different it will be severely damaged in its
vulnerable seedling stage where loss of leaf area can mean death.
    Only two *Passiflora* species known to me could possibly be
added to an equilibrium community in spite of leaf shape. Interestingly these are two of the most wide-spread species in Central
America. One, *Passiflora adenopoda* is unaffected by most
Heliconius because of its cuticular trichomes which kill larvae
(Gilbert, 1971). The second, *Passiflora serratifolia,* is, in the
insectary, a favorite oviposition plant for many species in the
genus. In fact, it is the "super-optimal" oviposition stimulus
for all *Heliconius melpomene* races which do not overlap its range.
P. serratifolia possess a chemical in the young leaves which,
though not detected by ovipositing females, kills all young larvae
(Gilbert, unpublished data).

Leaf shape and *Heliconius* diversity:

    It was pointed out earlier that for reasons not adequately
understood, the number of *Passiflora* species in an area predict
reasonably well the number of *Heliconius*. If, as hypothesized,
the degree of shape discrimination by the butterflies puts an
upper limit on the number of species of *Passiflora* which can be
packed into a community, then it follows that *Heliconius* can
indirectly limit their own local diversity.

CONCLUSION

The study of animal-plant coevolution has generally focused upon the interaction between plant and herbivore in isolation from concomitant interactions with other relevant members of the community. In this paper I have developed a conceptual model of the ecological and evolutionary feedback occurring within a food-web of interacting species, which is greatly amplified and accelerated by a mutualistic interaction.

The *Heliconius-Passiflora-Anguria* system is in many ways a special case which I have greatly oversimplified. For instance, few insects outside of the Lepidoptera interact with plants as both mutualist and herbivore. Even within the Lepidoptera, *Heliconius* may represent the most elaborate instance of this life-history type. However, since so little is known of the adult nutrition of most lepidopterans, the uniqueness of these butterflies remains an open question.

It should also be made clear that the model is derived from experiments and observations on just a fraction of the many species of *Heliconius*, *Anguria*, and *Passiflora*. I have knowingly ignored potentially important organisms such as the pollinators of *Passiflora*, other pollen plants of *Heliconius*, other pollinators, and herbivores of *Anguria*, to name a few.

Further research will almost certainly modify the details of Figure 1. However, I believe this somewhat specialized model illustrates some general ideas about the significance of mutualism in communities which deserve further testing by students of community ecology:

1. Animal-plant mutualisms depend upon and help determine the behavioral sophistication of the animal involved.
2. The behavioral capabilities of animals in a food-web strongly influence the local diversity attained by various taxa in the web as well as other emergent properties such as phenologies and patterns of morphological diversity among the plants. (Visual search for plants by host-specific herbivores may help generate morphological divergence within, and convergence among, chemically distinct groups of plants in a community.)
3. The importance of mutualism in terrestrial communities is greatest: a) where the animals involved have strong trophic ties to several distinct groups of organisms, and b) in the humid tropics where seasonal restraints on extended reproductive longevity and life history variety are lacking.
4. Since mutualistic interactions between animals and plants affect a community primarily by indirect pathways, their

consequences are less conspicuous and more difficult to
quantify than are those of predator-prey and competitive
interactions. In fact, in the case considered, the
mutualism determines the rules for these latter interactions.

SUMMARY

Heliconius butterflies have coevolved with two groups of
plants: *Passiflora* the larval hosts, and *Anguria/Gurania* which
are the primary pollen and nectar sources for the adults. As a
result of increased reproductive longevity, due to mutualistic
interaction with the cucurbit vines, *Heliconius* have evolved
highly sophisticated behavior patterns which include various forms
of learning and memory. Many features of the individuals and
populations as well as emergent properties of this coevolved sub-
community are casually linked either by direct or indirect pathways to the mutualistic interaction. It is suggested that animal-plant
mutualism can be a major factor in generating patterns of animal
and plant diversity.

NOTES

1. Baker (1963) was one of the first to recognize the need to
consider community level consequences of animal-plant mutualisms.
Recently such interactions have been discussed in relation to plant
spacing (Janzen, 1970, 1971; Heinrich and Raven, 1972), flowering
phenology (Janzen, 1967; Heinrich and Raven, 1972) and to community
diversity (Janzen, 1966, 1971). In one of the few theoretical considerations of the impact of mutualism on communities, May (1973)
concludes that in small model ecosystems mutualistic interactions
have a destabilizing influence. But mutualism as May treats it
has neither time lag nor inequality between the benefits exchanged
by species involved. It is consequently not representative of
most plant-animal mutualism. In contrast to well studied predator-
prey and competition interactions, mutualism has been almost
totally ignored by the model-makers of ecology.

2. *Gurania*, a cucurbit genus very closely related to *Anguria*,
contains several species which are primarily *Heliconius* pollinated
in addition to its many strictly hummingbird pollinated species.
For simplicity, I will occasionally lump all *Anguria* and *Gurania*
visited by *Heliconius* under "*Anguria*".

3. In none of 10 neotropical wet-forest sites ranging from
Trinidad to Mexico have I found more than 10 coexisting *Passiflora*
species. W. W. Benson (personal communication) has obtained a
similar impression in Central and South America.

4. *Heliconius*, in contrast to other tropical insect taxa, are
better researched than most or all similar temperate zone groups.
This fortunate state of affairs is due largely to two decades of
heliconiine research supported or encouraged by the William Beebe
Tropical Research Station near Arima, Trinidad. Much of the important literature on *Heliconius* biology has been reviewed elsewhere
(Ehrlich and Gilbert, 1973).

5. Preliminary experiments (Gilbert, unpublished data) indicate
that *Heliconius* can be conditioned to discriminate leaf models by
shape when associated with a nectar reward, but further work is
required to establish similar learning ability to be associated
with oviposition rewards.

6. One of the *Heliconius* stocks at Austin has now persisted in
captivity for 4-1/2 years. Several others are 2-3 years old. All
attempts to maintain other groups of butterfly for long periods
have failed. *Heliconius hecale* was maintained in a windowless
laboratory for three months. Visits to artificial nectar, mating,
roosting and oviposition on cuttings of host all proceeded quite
normally. The tops of tall burets and ring stand supports were
favored daytime perches!

7. For example: *Papilio zelicaon*, *Limenitis bredowii*, *Cethosia
cyane*, *Euphydryas editha*, *Actinote* spp., *Eumenus debora*, *Battus
philenor* (but some *Parides* do well).

8. This statement is based on my own visual comparisons of many
butterflies and should be verified by actual measurements.

9. Recently, I have found that male *Heliconius charitonia* actually
rape the female pupa (Gilbert, in preparation) as a routine mating
procedure.

10. Two frequently observed behaviors of egg-laying female butterflies indicates the use of chemical cues in recognition of
suitable larval host plant. The first, antennal tapping, presumably relates to olfactory recognition (Minnich, 1924); the
second, drumming the leaf with fore legs (Vaidya, 1956) is more
than likely a direct tasting of the leaf juices (Calvert, 1974).

11. *Heliconius charitonius* Linn., *H. cydno* Doubleday, *H. erato*
Linn., *H. ethilla* Godart, *H. hecale* Fabr., *H. melpomene* Linn.

12. *Anguria triphylla* Miq., *A. umbrosa* Kunth., *A. limonense*
Pittier, *A. warcewiczii* Hook., *Gurania spinulosa* Poepp. et Endl.,
G. costaricense Cogn., *G. levyana* Cogn.

13. Microkjeldahl analysis indicates that a large (no. 3) load of
Anguria pollen (dry wt. 0.7 mg) contains .028 mg nitrogen. Since
one .50 mg (wet wt.) *Heliconius* egg contains about .006 mg nitrogen, such a pollen load contains the nitrogen equivalent of about
five eggs (Gilbert and Norris, unpublished data).

14. We now know from radioisotope labelling experiments that
nitrogenous compounds in the spermatophore contribute directly to
egg production and may be of greatest importance in the first few
days of oviposition, before the female has located pollen sources
(Gilbert and Sumners, in preparation). In the insectary, males
seem to collect pollen more vigorously and successfully on the day
following a mating.

15. For instance, in one year (November 1971 - November 1972)
under greenhouse conditions an *Anguria umbrosa* male produced
10,000 flowers, the equivalent of 145 gm of dry sucrose and 20 gm
of pollen (Gilbert, unpublished data). As this paper is delivered
(August, 1973), this plant is still in flower.

16. *Anguria* pollen measures about 80 microns in diameter. It is
interesting that rubiaceous plants which attract *Heliconius*, such
as *Palicorea* and *Cephaelis* have unusually large (80-100 micron)
pollen for their family.

17. A female which lives for 4-6 months and routinely revisits
the same passion vines will quite likely encounter several generations of her own offspring emerging within her home range. If, as
limited evidence indicates, young individuals follow older butterflies, the chances of families roosting together would be high.
Allozyme studies of roosting groups would be of singular interest.

18. Kleber (1935) long ago demonstrated daily rhythms in nectar
flow (and nectar concentrations) among numerous European plants,
and partitioning of pollinators on a seasonal basis is documented
(Mosquin, 1971).

19. Ovipositing female butterflies occasionally deposit eggs on
inappropriate host plants (e.g. Singer, 1971). Such behavior often
results from a rarity of the primary or favored host, possibly due
to defoliation. I have observed insectary *Heliconius* oviposit on
cucurbit vines which resemble the appropriate host. Lesser mistakes, i.e. eggs-laid on the wrong *Passiflora*, are not unlikely
and are indicated by eggs found in the field on incorrect *Passiflora* (Alexander, 1961a).

20. W. W. Benson (personal communication) has independently
noticed the egg mimics on several other species of *Passiflora*.

ACKNOWLEDGEMENTS

    I owe particular thanks to Paul R. Ehrlich who provided
support (NSF-BG19686 and GB22853X) and collaboration with the
first study in Trinidad and all other persons associated with that
work. Equal thanks go to the Organization for Tropical Studies in
Costa Rica, where, through use of facilities and interaction with
colleagues, I have benefited immeasurably. Woody Benson, Keith
Brown and John Turner have provided many helpful letters and discussions and will undoubtedly be able to provide other, perhaps
better, insights into the biology of *Heliconius*. I hope I have
accurately acknowledged their work in the text. Other persons providing helpful discussions, criticisms of the paper or general
inspiration are Martha Condon, Dan Janzen, Gordon Orians, Mike
Singer, John Smiley and Joan Strassman. To these and many others
who have provided useful comments at seminars I am grateful. Much
of the research discussed above is supported by NSF grant GB4074X-P.

LITERATURE CITED

Alexander, A. J. 1961a. A study of the biology and behavior of
    the caterpillars and emerging butterflies of the subfamily
    Heliconiinae in Trinidad, West Indies. Part I. Some aspects
    of larval behavior. Zoologica 46:1-24.

Alexander, A. J. 1961b. A study of the biology and behavior of
    the caterpillars and emerging butterflies of the subfamily
    Heliconiinae in Trinidad, West Indies. Part II. Molting,
    and the behavior of pupae and emerging adults. Zoologica 46:
    105-124.

Baker, H. G. 1963. Evolutionary mechanisms in pollination biology.
    Science 139:877-883.

Benson, W. W. 1971. Evidence for the evolution of unpalatability
    through kin selection in Heliconiinae. Amer. Natur. 105:213-
    226.

Benson, W. W. 1972. Natural selection for mullerian mimicry in
    Heliconius erato in Costa Rica. Science 176:936-939.

Bernstein, S. and Bernstein, R. A. 1969. Relationships between
    foraging efficiency and the size of the head and component
    brain and sensory structures in the red wood ant. Brain
    Research 16:85-104.

Brower, L. P., Brower, J. V. Z. and Collins, C. T. 1963. Experimental studies of mimicry. 7. Relative palatability and mullerian mimicry among neotropical butterflies of the subfamily Heliconiinae. Zoologica 48:65-84.

Brower, L. P. and Brower, J. V. Z. 1964. Birds, butterflies and plant poisons: a study in ecological chemistry. Zoologica 49:137-158.

Brown, K. S. and Mielke, O. H. H. 1972. The Heliconians of Brazil (Lepidoptera: Nymphalidae). Part II. Introduction and general comments, with a supplementary revision of the tribe. Zoologica 57:1-40.

Calvert, W. H. 1974. The external morphology of foretarsal receptors involved with host discrimination by the nymphalid butterfly, *Chlosyne lacinia*. Ann. Ent. Soc. Amer. (in press).

Cheesman, E. E. 1940. Cucurbitaceae (Passiflorales). *In* Flora of Trinidad and Tobago, Vol. I. pp. 436-448.

Cogniaux, A. 1924. Cucurbitaceae - Fevilleae et Melothrieae. *In* Engler Ed. Pflanzenreich IV 275:178-230.

Crane, J. 1957. Imaginal behavior in butterflies of the family Heliconiidae: changing social patterns and irrevelant actions. Zoologica 42:135-145.

Dahlgren, R. 1971. Multiple similarity of leaf between two general of Cape plants, *Cliffortia* L. (Rosaceae) and *Aspalathus* L. (Tabaceae). Botaniska Notiser 124:292-304.

Edwards, W. H. 1881. On certain habits of *Heliconia charitonia* Linn., a species of butterfly found in Florida. Papilio 1: 209-215.

Ehrlich, P. R. and Raven, P. H. 1965. Butterflies and plants: A study in coevolution. Evolution 8:586-608.

Ehrlich, P. R. and Gilbert, L. E. 1973. Population structure and dynamics of the tropical butterfly *Heliconius ethilla*. Biotropica 5:69-82.

Emsley, M. G. 1965. Speciation in *Heliconius* (Lep. Nymphalidae): morphology and geographic distribution. Zoologica 50:191-254.

Fisher, R. A. 1930. The Genetical Theory of Natural Selection. Oxford University Press.

Futuyma, D. J. 1973. Community structure and stability in constant environments. Amer. Natur. 107:443-446.

Gilbert, L. E. 1971a. Butterfly-plant coevolution: has *Passiflora adenopoda* won the selectional race with heliconiine butterflies? Science 172:585-586.

Gilbert, L. E. 1971b. Distribution and abundance of resources of factors which determine population structure in butterflies. Ph.D. Thesis, Stanford University.

Gilbert, L. E. 1972. Pollen feeding and reproductive biology of *Heliconius* butterflies. Proc. Nat. Acad. Sci. 69:1403-1407.

Gilbert, L. E. and Singer, M. C. 1973. Dispersal and gene flow in a butterfly species. Amer. Natur. 107:58-72.

Heinrich, B. and Raven, P. 1972. Energetics and pollination ecology. Science 176:597-602.

Janzen, D. H. 1966. Coevolution of mutualism between ants and acacias in Central America. Evolution 20:249-275.

Janzen, D. H. 1967. Synchronization of sexual reproduction of trees within the dry season in Central America. Evolution 21:620-637.

Janzen, D. H. 1970. Herbivores and the number of tree species in tropical forests. Amer. Natur. 104:501-528.

Janzen, D. H. 1971. Euglossine bees as long-distance pollinators of tropical plants. Science 171:203-205.

Killip, E. P. 1938. The american species of Passifloraceae. Publ. Field. Mus. Nat. Hist. (Bot) 19:2-613.

Kleber, E. 1935. Hat das Zeitgedachtnis der Bienen Biologische Bedeutung? Z. vergl. Physiol. 22:221-262.

Koltermann, R. 1971. 24-Std-Periodik in der Langzeiterinnerung an Duft-und Farbsignale bei der Honigbiene. Z. vergl. Physiol. 75:49-68.

Levins, R. 1975. Evolution in communities near equilibrium. *In* Ecology and Evolution of Communities. Belknap Press, Harvard University. (in press).

Lloyd, D. G. 1973. Sex ratios in sexually dimorphic umbelliferae. Heredity 31:239-249.

May, R. M. 1973. Qualitative stability in model ecosystems. Ecology 54:638-641.

Minnich, D. E. 1924. The olfactory sense of the cabbage butterfly, *Pieris rapae* Linn., an experimental study. J. exp. Zool. 39:339-359.

Mosquin, T. 1971. Competition for pollinators as a stimulus for the evolution of flowering time. Oikos 22:398-402.

Poulton, E. B. 1931. The gregarious sleeping habits of *Heliconius charitonius* L. Proc. Roy. Entomol. Soc. London 6:4-10.

Remington, C. L. 1963. Historical backgrounds of mimicry. Proc. XVI International Congress of Zoology, Vol. 4:145-149.

Richards, P. W. 1964. The Tropical Rain Forest. Cambridge University Press.

Singer, M. C. 1971. Evolution of food-plant preference in the butterfly *Euphydryas editha*. Evolution 25:283-389.

Swihart, C. A. 1971. Colour discrimination by the butterfly *Heliconius charitonius* Linn. Anim. Behav. 19:156-164.

Swihart, S. L. 1963. The electroretinogram of *Heliconius erato* (lepidoptera) and its possible relation to established behavior patterns. Zoologica 48:155-165.

Swihart, S. L. 1964. The nature of the electroretinogram of a tropical butterfly. J. Ins. Physiol. 10:547-562.

Swihart, S. L. 1967. Hearing in butterflies (Nymphalidae: *Heliconius*, *Ageronia*). J. Ins. Physiol. 13:469-476.

Turner, J. R. G. 1971. Experiments on the demography of tropical butterflies II. Longevity and home-range behavior in *Heliconius erato*. Biotropica 3:21-31.

Vaidya, V. G. 1956. On the phenomenon of drumming in egg-laying female butterflies. J. Bombay Nat. Hist. Soc. 54:216-217.

Vaidya, V. G. 1969b. Form perception in *Papilio democleus* L.
    (Papilionidae, Lepidoptera). Behavior 33:212-221.

Williamson, M. 1972. The analysis of biological populations.
    Edward Arnold, London.

Wickler, W. 1968. Mimicry in plants and animals. McGraw-Hill,
    New York.